Charles Ottley Groom Napier

Natural History Rambles

Charles Ottley Groom Napier

Natural History Rambles

ISBN/EAN: 9783337026424

Printed in Europe, USA, Canada, Australia, Japan

Cover: Foto ©berggeist007 / pixelio.de

More available books at **www.hansebooks.com**

NATURAL HISTORY RAMBLES.

LAKES AND RIVERS.

BY

CHARLES OTTLEY GROOM NAPIER, F.G.S.

AUTHOR OF

"THE FOOD, USE, AND BEAUTY OF BRITISH BIRDS,"
"MISCELLANEA ANTHROPOLOGICA," "THE BOOK OF NATURE
AND THE BOOK OF MAN," ETC.

PUBLISHED UNDER THE DIRECTION OF
THE COMMITTEE OF GENERAL LITERATURE AND EDUCATION
APPOINTED BY THE SOCIETY FOR PROMOTING
CHRISTIAN KNOWLEDGE.

SOCIETY FOR PROMOTING CHRISTIAN KNOWLEDGE.
LONDON: 77, GREAT QUEEN ST., LINCOLN'S-INN FIELDS, W.C.;
4, ROYAL EXCHANGE E.C.; 48, PICCADILLY, W.
AND BY ALL BOOKSELLERS.
NEW YORK: POTT, YOUNG & CO.

1879.

CONTENTS.

CHAPTER		PAGE
	INTRODUCTION	7
I.	MAMMALS.—THE WATER-SHREW, THE OTTER, AND THE WATER-VOLE	11
II.	AQUATIC BIRDS.—RAPACES, DIPPER, KINGFISHER, TITS, AND WAGTAILS	24
III.	A RAMBLE AMONG THE MARSH BIRDS OF SUSSEX	48
IV.	AQUATIC AND MARSH BIRDS.—HERONS, PLOVERS, STORKS, SANDPIPERS, SNIPES, RAILS, DIVERS, GREBES, GEESE, DUCKS...	62
V.	AMPHIBIA.—FROGS, TOADS, AND NEWTS ...	120
VI.	FRESH-WATER FISHES	129
VII.	SOME OF THE TYPICAL RIVERS AND THEIR FISHERIES	158

CHAPTER		PAGE
VIII.	OUR AQUATIC MOLLUSCA.—SPHÆRIUM, PISIDIUM, UNIO, PEARL MUSSEL, ANODONTA, DREISSENA, NERITINA, PALUDINA, VALVATA, PLANORBIS, LIMNEA	191
IX.	SIX MONTHS OF BROOK AND POND LIFE IN SUSSEX	220

LAKES AND RIVERS.

INTRODUCTION.

THE fresh-water system of Great Britain affords the naturalist a field for observation too extensive to be exhausted within the limits of a small book. Hence the necessity of making some selection, so as to show what is most characteristic, to awaken observation, and entice on to the perusal of more compendious monographs, which those who pursue a speciality among any of the departments of fresh-water life will require. Among the books best known to the author, where further information can be found, are, Bell's "British Quadrupeds," Macgillivray's "British Quadrupeds," Yarrell's "British Birds," Macgillivray's "British Birds," Bell's "British Reptiles," Hewitson's "Eggs of British Birds," Couch's, Yarrell's, and Buckland's works on British Fishes, and Curtis's illustrations of British Entomology. I do not, however, know of any really satisfactory work treating of British aquatic insects, the information obtainable being scattered over perhaps a hundred different books and articles, each of which gives small contributions to the subject.

The fresh-water shells and their inhabitants are well treated in J. Gwyn Jeffreys' "British Conchology." There are a few little books on pond and freshwater life in its lower forms, but I do not know any really important works which exhaust the subject.

The British flowering plants and *algæ* have been well treated in the various editions of Sowerby's "English Botany," of which the most complete is undoubtedly that edited by Mr. Boswell Syme and Mrs. Lankester, although the *algæ* and other lower cryptogamic forms are wanting. This little work comprises the observations of the author during a quarter of a century, guided by the books above specified, and many others which space is wanting to mention.

Among the subjects will be found personal observations on the habits of the water-shrew, otter, and water-vole, observations on our water birds, amphibia, fresh-water fish, insects, mollusca, lower forms of animal life, and aquatic botany. The fauna and flora of the banks of rivers and lakes will likewise be treated of, as well as those of marshes; and occasional reference will be made to the marine animals which visit rivers, but space will not admit of their minute description. The results of my examination of fish attacked by the too-prevalent fungus *Caprolignea ferax*, for the Salmon Fisheries Report laid before Parliament in 1878, will also be given.

Although the work is for the most part systematically arranged, yet for the sake of variety a ramble among the marshes of Sussex, descriptive of the nesting habits of the marsh birds, and six months of aquatic life in Sussex will be given. An account of a few of our most typical rivers and their fisheries has also been included. Figures of most of the freshwater mammals, birds, and fish are given.

CHAPTER I.

MAMMALS.—THE WATER-SHREW, THE OTTER, AND THE WATER-VOLE.

WE shall begin our examination of the fauna of British lakes and rivers with the Mammals, as being the most highly-organized, and therefore meriting

THE WATER-SHREW.

most attention. Among these the order *Insectivora* occupies a prominent place, and there are several familiar examples to be encountered in nearly every stream. There is, for instance, the Water-shrew, whose habits are worthy of careful attention. The Water-shrew (*Sorex fodiens*, Jennings) is somewhat larger than the common shrew, from which it is distinguished by being darker on its upper portions, which are separated by an abrupt line from the lighter or lower part : its length, including the tail, is $5\frac{1}{2}$ inches. The nose is flattened, and it has very short round ears, with three internal lobes; the tail is nearly square; the fur is very velvety, and varies in colour from black to dark brown; the incisor teeth are bright red-brown at the tip. This little animal is widely distributed in England and Scotland; it makes a burrow on the banks of a stream or pond, the mouth of which is easy of access from the water; it dives with facility and swims with great rapidity, making but a slight ripple on the water, so as not to disturb the insects which form its food. I have dissected four, and found their stomachs to contain nothing but the remains of water-beetles, dragon-flies, and their larvæ. About twenty years ago, in Sussex, I had a good opportunity of examining these animals. I noticed a family of four, two half-grown, leave a burrow, the entrance of which would about admit the thumb. The old ones taking to the water, swam down the stream, while the young followed each other in diving. There is wonderful benefit to the naturalist in being able to move without noise, and observe without being observed. If a man goes to a marsh and remains for some time perfectly still, he will in most cases get an insight into the habits of animals which the talkative or bustling observer cannot do. This was the experience of Audubon, of Waterton, and of others whom I have known, of less note. Remembering these precepts, I threw myself full-length on the grass and waited for a quarter of an hour, by which time the whole family of water-shrews had returned laden to their hole, each carrying a water-beetle. This was a small experience, but surely one worth a quarter of an hour's silent observation.

Mr. Dovaston, in the second volume of the "Magazine of Natural History," thus describes these animals:—
"On a delicious evening far in April, 1825, a little before sunset, strolling in my orchard beside a pool, and looking into the clear water for insects I expected about that time to come out, I was surprised by seeing what I momentarily imagined to be a *Dytiscus marginalis*, or some very large beetle, dart with rapid motion, and suddenly disappear. Laying myself down cautiously and motionless on the grass, I soon, to my delight and wonder, observed it was a mouse. I repeatedly marked it glide from the bank under water and bury itself in the mass of leaves at the bottom; I mean the leaves that had fallen off the trees in autumn, and which lay very thick in the mud. It very shortly returned, and entered the bank, occasionally putting its long sharp nose out of the water, and paddling close to the edge. This it repeated at very frequent intervals from place to place, seldom going more than two yards from the side, and always returning in about half a minute. I presume it sought and obtained some insect or other food among the rubbish and leaves, and retired to consume it. Sometimes it would run a little on the surface, and sometimes timidly and hastily come ashore, but with the greatest caution, and instantly plunge in again. During the whole sweet spring of that fine year I constantly

visited my new acquaintance. When under water he looks gray, on account of the pearly cluster of minute air-bubbles that adhere to his fur and bespangle him all over. He swims very rapidly, and though he appears to *dart*, his very nimble wriggle is clearly discernible."

As an example of the order *Carnivora* we give the Otter (*Lutra vulgaris*), the most important of our aquatic mammalia. It is allied to the weasels, among which Linnæus classed it. It is found from $3\frac{1}{2}$ to 4 feet in length, and varies in colour according to age and locality, from grayish-brown to chocolate on the back, while the under parts of the neck and breast are grayish or brownish white. The head is very wide and flat, the muzzle short, broad, and round, with a thick upper lip and a flat nose ; the neck is very thick, almost as much so as the thorax. The body is very much elongated and flexible, having almost a snake-like movement ; its feet have five sharp prehensile claws, which are webbed somewhat like those of a seal; the tail is long, thick, and muscular, and is most important in the economy of the animal, as facilitating its steering and swimming, for it acts both as a rudder and as a scull, a peculiarity connecting it with the Fishes, although its tail is not nearly of the importance that a fish's tail is. It would be possible for an otter to live without its tail, not so a fish. The teeth of the Otter are very powerful, which any one finds to his cost who has the misfortune to be bitten by the animal ; they seize with the ferocity and tenacity of a ferret, but from their much greater size with vastly more power. The fur of the Otter is extremely serviceable, comparable to beaver and sealskin in many respects, although inferior in lustre and richness of colour; but brings a very high price when it is in a condition —a thing rare in England—suitable for making high class trimmings for ladies' jackets. The longer hairs are somewhat coarse, and these in the manufacturing process are removed, so that the fur is left more silky than anything the human loom can weave. The Otter feeds almost entirely on fishes, in rivers, but has no objection to crustacean dainties at sea. It swims with great rapidity on the surface of the water and can readily dive below its surface; but for these rapid movements it could not at all times obtain a plentiful supply of fish. It brings its prey to land to devour it, for it could not eat it in the water without swallowing a great quantity of that liquid. It is this habit that renders it a matter of but little difficulty to shoot otters, if the sportsman knows the time to be still.

The Otter is dainty, eating only the more juicy and fleshy portions of the fish. Its lair is often thickly strewed with half-devoured fish. A Mr. Morgan O'Brian, a gamekeeper in the county of Kerry, wrote to me many years ago an interesting letter, in which he described the habits of the Otter in his county. There was one remarkable lair on the banks of the Feale which was occupied by a family of five otters; it was situated on a peninsula which ran into the river; the stronghold was under the root of a large willow tree, very difficult of access to dogs: there had always been otters there in his knowledge, which extended over twenty years. For five successive summers he used to visit the peninsula daily, the first thing in the morning.

Large numbers of fish were lying about, sometimes alive. Among the fish were trout, large eels, and occasionally salmon; he got on the average a dozen pounds of fish weekly from this source. One summer he captured three young otters from this lair; two he gave away, but the third he trained to fish in the river at a point where the fish were unusually large and abundant. This otter became remarkably tame and intelligent, and was for several years his humble servant, when it was unfortunately killed by the bull-dog of a visitor, not, however, before biting a large piece out of the cheek of its canine assailant. Mr. Duncan says of the Otter, "While eating, it holds its prey in its fore-feet: or, if small, it secures it between them, and commencing at the shoulders, devours the fish downwards, leaving the head and tail." While thus occupied, it is sometimes visited by gulls and hooded crows, which, however, do not venture to attack it, but wait until it has finished its meal, contenting themselves with the remnants. It is alleged that it destroys great quantities of salmon; which may be the case when it inhabits rivers or estuaries in which that fish is abundant, but in the open sea it feeds on a variety of fishes. Along the coast it finds generally a safe retreat in coves of which the upper part is filled with blocks of rock, or beneath large stones; but in rivers and lakes it seeks refuge among the roots of trees, or burrows a hole for itself in the bank. Although principally piscivorous, it has been known to attack young domestic animals, and I found the stomach of one killed in June filled with a curious collection of larvæ and earthworms.

The female is said by Mr. Bell to go with young nine weeks, and to produce from three to five young ones in March or April. I cannot confirm or refute these assertions, but I have examined an individual so young as to be still sucking, without the lower incisors, and only 20½ inches in length, which was killed near the top of the river Don on the 25th of November. At this early age the head is round and flattened, the eyes placed so near the nostrils that three lines drawn, the first from one eye to the other, the second and third from each eye to the middle joint between the nostrils, form an equilateral triangle. The hair of the lips and face is shorter and stronger than elsewhere, and of a grayish colour; under the nostrils are two nearly contiguous yellowish spots; the claws are very acute, the tail proportionately shorter and depressedly conical; the general colour sooty-brown.

A gentleman residing in Berneray, in the Outer Hebrides, had an otter that supplied itself with food, and regularly returned to the house. Mr. McDiarmid, in his amusing "Sketches from Nature," gives an account of several domesticated otters, one of which belonged to a poor widow, which "when led forth plunged into the Urr, or the neighbouring burns, and brought out all the fish it could find." Another, kept at Crosbie House, Wigtonshire, evinced a great fondness for gooseberries, fondled about its keeper's feet like a pup or kitten, and even seemed inclined to salute her cheek, when permitted to carry its freedoms so far. A third, belonging to Mr. Monteith, of Carstairs, was also very tame; and though it frequently stole away at night to fish by the pale light of the moon and associate with its kindred by the river-side, its master of course was too generous to find any fault with its peculiar mode of spending its evening hours. In the morning it was always at its post in the kennel, and no animal understood better the secret of keeping its own side of the house. Indeed, its pugnacity in this respect gave it a great lift in the favour of the gamekeeper, who talked of its feats wherever he went, and averred, besides, that if the best cur that ever ran "only daured to grin" at his *protégé*, it would soon make its teeth meet through him. To mankind, however, it was much more civil, and allowed itself to be gently lifted by the tail, though it objected to any interference with its snout, which is probably with it the seat of honour. Otter-hunting was much more frequently pursued a few generations ago than it is at the present day. An instance proving that otters are able to catch even the largest of the inland-water fish has been communicated to the *German Fishing Gazette* as having happened in Norway:—"The fresh remains of an otter-meal were discovered a few days ago upon the banks of the Lardalscly, in Norway, consisting of the head and tail-end of a salmon. The weight of the head was six pounds, and that of the tail-end up to the lowest point of the dorsal fin, twenty English pounds. The marks of the fore-paws of the otter upon the tail-end of the salmon indicated clearly that the otter must have caught hold of the fish at the tail, and that he had let himself be dragged along by the salmon until the latter's strength had been entirely exhausted, when it fell an easy prey to its enemy. To judge from the proportions of the discovered remains, the total weight of the salmon could not be estimated at less than fifty pounds; the otter having, therefore, in one single meal devoured some twenty-five pounds of fish-flesh."—"J.," in *Land and Water*, Nov. 9, 1878.

The only aquatic example of the order *Rodentia* in Britain is the *Arvicola amphibius,* Water-vole or Water-rat, a familiar animal in most country places in the

WATER-RAT.

vicinity of water. The Voles are less elongated than the Rats and Mice properly speaking, and have broader heads, shorter ears and tails, and are, in my opinion, more attractive. The body of the brown Water-vole, which is usually called the Water-rat, is rather stout, and the neck and head are short; the latter is round and convex on the top;

the legs are small, and the tail is slender and not very long. The eyes are small, and the nostrils are placed sideways on the snout, which is short. The ears are scarcely visible, being covered in the fur; their inside is bare, but the edges of the ears have a fringe of fine hair; the meatus is large, and can be completely covered by its operculum, which has a thickened margin. The fore-feet have each five toes of unequal length, the claws are much bent in. The sole of the foot is bare, with papillæ-like protuberances opposite the toes. The hind feet have also five toes, the first of which is short, and the next three equal in length, the fifth much shorter. The claws are longer and stronger than those of the fore-feet. The sole is bare to the tarsal joint, and there are also tubercles at the roots of the toes, and one between the third and fourth. The tail is covered with short hairs, and is round and tapering. "The upper incisors are of moderate size, slightly curved and chisel-shaped, and close together; those of the lower jaw are long, curved, sub-cylindrical, rounded, but naturally sloped." Inside the mouth, where the lips meet, there is a tuft of white shining hairs, growing out of a hard skin. The animal has three grinders on the upper and under jaws; the sides of these teeth are grooved with an irregular appearance. The palate has several rows of tubercles on its surface, which in the inner row become very small, and terminate near the first grinder. The stomach is proportionately large, and has a middle contraction, which divides it into two. The intestine is about 5 feet long, and the cæcum is very large. The pile is very long and thick; there are two sets of hairs of different lengths; the general appearance is soft and rather silky. The colour of the longer hairs is dark brown, or a mixture of brownish-black and reddish-brown underneath. The belly and lower parts of the animal, as well as the sides of the head, are lighter in colour. Those parts of the hairs which are usually not visible, are of a bluish-black or bluish-gray; the tips of the shorter hairs are throughout of a reddish-brown. The teeth are brownish-yellow, the nose is dusky, the soles of the feet pale flesh-colour, and the claws yellowish:

Length of male	12 in.;	of female	12 in.	10 lin.
Of head	,,	1 in. 10 lin.	,,	1	10
Of head and body	8	0	,,	8	6
Of tail	4	0	,,	3	9
Of fore-foot	0	9	,,	0	9
Of hind-foot	1	3	,,	1	3

The Water-vole is common in Britain, but it is not found in some parts of the North of Scotland. It lives in the banks of rivers, small streams, canals, or ponds. It scoops out long, crooked burrows for itself, and frequently forms a passage under the water, so that when it is swimming, and is in danger from an enemy, it can get to its hole without coming to the surface of the water, as it is an excellent diver. It lives on vegetables, and usually feeds in the morning and evening. It deposits a store of provisions for the winter, sometimes hoarding potatoes; it is not, however, torpid during the cold season, though it remains in its hole in time of snow. It makes a nest of dried grass or vegetable remains, where it deposits five or six young in the beginning of summer. This rat does not frequent human habitations, and is not generally destructive, unless its burrows can be said to spoil the banks of canals. Like other species of rats and mice, this vole is sometimes found as an albino. Mr. Macgillivray describes a variety of it which is black, and which he names *Arvicola ater*. He at one time thought it a distinct species, but afterwards seemed undecided about this, the skeletons being similar, as well as the fur, except in colour. The chief difference, he says, is in the size of the skull, as the black variety is generally smaller than the brown. Similar differences, however, frequently occur in animals of undoubtedly the same species.

Early in the spring of 1855, I dug out the burrow of a water-vole, and was surprised to find at the further extremity a cavity of about a foot in diameter, containing a quantity of fragments of carrots and potatoes, sufficient to fill a peck measure. This was undoubtedly its winter store of provisions. This food had been gathered from a large potato and carrot-bed in the vicinity. On pointing out my discovery to the owner of the garden, he said his losses had been that winter very serious, owing to the ravages of these animals, and said he had brought both dogs and cats down to the stream to hunt for them, but they were too wary to be often caught. White, of Selborne, says:—"As a neighbour was lately ploughing in a dry, chalky field far removed from any water, he turned out a water-rat that was curiously laid up in an hybernaculum, artificially formed of grass and leaves. At one burrow lay above a gallon of potatoes, regularly stowed, on which it was to have supported itself for the winter. But the difficulty with me is how this *amphibius mus* came to fix its winter station at such a distance from the water. Was it determined in its choice of that place by the mere accident of finding the potatoes which were planted there; or is it the constant practice of the aquatic rat to forsake the neighbourhood of the water in the colder months?" (White's "Nat. Hist. of Selborne," Bell's ed. pp. 76, 77.) In Buckland's edition of White, page 443, an albino water-rat is mentioned as being caught when out fishing; it had pink eyes, and is of extreme rarity. The Water-rat is often devoured by the Pike, and perhaps also by the largest trouts. An interesting writer in Cassell's "Natural History," vol. i. p. 310, says:— "We have seen water-rats cross a wide meadow, climb the stalks of the dwarf beans, and after detaching the pods with their teeth, shell their contents in the most workmanlike manner. They will mount vines and feed on the grapes; and a friend informs us that on one occasion he saw a water-rat go up a ladder which was resting against a plum-tree and attack the fruit. If a garden is near the haunt of water-rats, it is necessary to watch narrowly for the holes underneath the walls, for they will burrow under the foundation with all the vigour of sappers and miners. Such is the cunning with which they will drive their shafts, that they will ascend beneath a stack of wood, a heap of stones, or any object which will conceal the passage by which they obtain an entrance. The Water-rat is a very clean animal in its habits. The flesh is said to be eaten by the French peasants on *maigre* days. It is found in most parts of Europe."

CHAPTER II.

AQUATIC BIRDS.—RAPACES, DIPPER, KINGFISHER, TITS, AND WAGTAILS.

THE Sea Eagle (*Aquila albicilla*), although usually a bird frequenting the cliffs of a few localities in remote parts of our sea-coasts, yet is known to haunt the vicinity of inland lakes and rivers. Mr. Thompson mentions it as frequenting the lakes of Killarney, and it used to breed on the Mourne mountains. Mr. Hewitson, quoting Mr. Wolley, says:—" The Sea Eagle generally makes its nest in the high cliffs of the coast, where it lives upon fish, guillemots, young herring, gulls, &c., but is also occasionally found breeding inland. In the former situation an eyrie which I visited two years in succession, and from which I took the egg which Mr. Hewitson figures, had nothing, but a very little heather, grass, and moss used in its construction. Two other nests which were carefully described to me were made principally of sea-weed, and were in such 'tremendous cliffs' that my informant's 'hair gets strong' when he thinks of them. In the Shetlands an inaccessible eyrie was pointed out to me on the extreme top of a stack, that is, a steep, detached rock; and I have seen another such stack on the North-east coast of Scotland, which was also said to have an eyrie at its summit. In inland situations the Sea Eagle is rare when compared with the Mountain Eagle (as the Golden Eagle is usually called), and it generally establishes itself upon a rock or islet in the middle of a loch. Here it builds upon the ground, or in a tree, a nest whose construction does not at all differ

WHITE-TAILED EAGLE.

from that of the other eagle, there being always in it a certain amount of *Luzula sylvatica*. The tree need by no means be a large one; I have seen two nests of different years in trees on separate islands in one loch, each only about four feet from the ground. I can, at this moment, call to mind nine instances where I know the localities of such island eyries. The old birds do not always calculate the depth of the water, as there is one place at least to which a man may wade; where swimming is necessary, it is often an affair of danger, as the birds will do their best to drown the enemy with their wings. In two spots I have seen large Scotch firs which have been formerly tenanted by sea eagles; one by the side of a loch, the other several miles away from any piece of water, in a sort of open wood of similar trees. The nest had been in a fork where three branches met, 20 feet high, and as in other cases the main trunk bore its weight. In one instance the crossed and nearly horizontal trunks of two small trees formed the support; one that I have already spoken of was in a small alder-tree, and had been repaired and often frequented by the eagles the season I saw it, yet a hooded crow had eggs in the upper branches, and wild geese and ducks were sitting in the deep moss and long heather within twenty yards. I have not myself met with an instance of the Golden Eagle breeding in a tree or in a sea-cliff, but on the other hand several cases of the sea eagles breeding in a rock inland, though not many miles away from the ocean: two such nests that I visited were in small rocks of easy access, in every respect like golden eagles', and in one the hen showed the same unwillingness to leave her eggs. In the summer of 1848 I took out of their nest, on a ledge of a perpendicular sea-cliff, two full-fledged eaglets, which now, nearly five years old, have not (March, 1853) acquired their complete adult plumage, though kept in a most congenial situation among the Derbyshire rocks. The eggs of the White-tailed Eagle are laid a week or fortnight later than those of the Golden Eagle: they are generally smaller."

The prevailing colour of the adult Sea Eagle is a dull brown, paler on the head and upper parts, the feathers of which are pointed or lance-shaped. When the bird is irritated or alarmed, it raises its feathers, which then give it a much lighter appearance, and if the sun be shining it looks bright, and almost white. The quills are blackish-brown with a purple-tinge; the shafts are pale. The upper tail-coverts and the tail itself are pure white. This is the full adult plumage, usually complete after the third moult, but in the case of the bird above mentioned, the tail was not perfectly white at the age of five years, and was not so until it had attained the age of seven years. The plumage of the young is of an amber-brown, grayer beneath, and the tips of the feathers paler, often white at the base. The tail is mottled brown and white. At each succeeding moult the tail becomes lighter; the colour of the beak becomes greenish, and the iris pale chestnut-brown, sometimes yellow. The shape of the Sea Eagle is more slouching than that of the Golden Eagle, and it is apt, when sitting at rest, to hang its feathers loosely, so as to appear untidy and sluggish.

The Osprey (*Pandion haliætus*), unlike the Sea Eagle, takes its prey from the water; its claws are very sharp and admirably adapted for grasping, being comparable to grappling-hooks, with the advantage of being capable of direction with the celerity and skill which no inanimate objects possess. Macgillivray says :—" The outer toe is versatile, and the lower surface of the whole foot, particularly on the raised pads

THE OSPREY.

which are placed under the joints, have the surface studded with hard and sharp pyramidal points, which enables them to hold their prey, however slippery, whether pierced by the claws or not. The claws are very strong, curved so as to form part of a circle, and remarkably sharp, while they exhibit another peculiarity in being perfectly rounded—a structure which will allow either an easy piercing or a quick retractation or loosening of the hold, if the prey should prove too weighty. The length of this bird is about 23 inches, the beak is much hooked and is of a bluish-black; it has long lanceolate feathers on the back of the head, which when raised look like a crest. The head and nape are white and brown. The posterior part of the auriculars, wings, and back, are dark-brown, but the tips of the feathers on the back and scapulars are paler. The wing-feathers are rather longer than those of the tail; they are white at the under part, and marked with brown. The tail is short and square the feathers are ample, the colour is brown with pale bars, the shafts and under parts are white." Mr. Wolley says:—"I have seen several nests of the Osprey upon the highest points of ruins, in and about lochs in Scotland, and several more upon small isolated rocks projecting out of the water. There is something in the general appearance of the nest which reminds one of nests of the wood-ants; it is usually in the form of a cone cut off at the top. The sticks project very slightly beyond the sides, and are built up with turf and other compact materials; the summit is of moss, very flat and even, and the cavity occupies a comparatively small part of it. I know no other nest at all like it. There was a nest for some years on the sloping trunk of a tree, which several persons have described to me. The birds are very constant year after year in returning to their old stations, and even after one or both birds have been killed in the previous season I have frequently seen individuals flying near the now-deserted eyrie."

Mr. Selby has seen the osprey on Loch Awe,—"where an eyrie is annually established upon the ruins of a castle near the southern extremity of the lake, and another in a similar situation nearly opposite the egress of the river Awe."

The Osprey breeds in May, and lays from two to four eggs, which have the ground-colour bluish or yellowish-white, variously spotted with burnt-sienna, light and dark and pale lilac; some are very richly marked with large blotches of a dark sienna. The size is that of a common domestic duck.

The Marsh Harrier (*Circus rufus*) is a bird of prey which frequents low and marshy grounds, and feeds on aquatic birds, quadrupeds, and frogs.

Gliding over the tops of the reeds, it pounces on its prey. The Marsh Harrier was once common in the fenny districts of England, but, through the drainage of these, it has become almost extirpated. The colour of this bird varies: sometimes the head and throat have much white, at other times the plumage, except the fore-head, hind-head, throat and sides of the mouth, are deep umber-brown with no light shade underneath; the parts excepted are yellowish-white. In another, the plumage is pale yellowish-red on the upper tail-covert and base of the tail-feathers. The remainder of the feathers are chiefly umber-brown. The quills are not tipped with white, and the white of the hind-head is pure, also above each eye. In the young the colour is more uniform and there is no yellowish-white about the head. In another variety the pale colour is sometimes all over the head, and a patch of the same on the scapulars. But the most usual is that of the first described. Mr. Selby mentions one which he kept which had "the throat, bastard wing, four quill-feathers, and outer tail-feathers white." The bill is bluish-black, the feet yellow, and the claws sharp, but not very strong. The usual length of the female is from 22 to 24 inches, and the male is several inches less. Mr. Gould, in his "Birds

THE MARSH HARRIER.

of Europe," states his opinion that most of the marsh harriers found in Britain are young birds, and that the species takes a long time to become perfectly mature, but that it breeds before being so.

Professor Alfred Newton says that the Marsh, Common, and Montague's harriers used to breed in the fens of Norfolk, but are now nearly extinct in these breeding-places. The eggs of all three species are of a pale blue colour, and are sometimes spotted with genuine blotches from the bird of an ochre colour, at other times are stained, like those of the Grebes, with matter from the claws of the bird or from the lining of the nest. The eggs of the Marsh Harrier are about the size of a small fowl's; those of the other species are considerably less. The nest of the Marsh Harrier is composed of bulrushes, reeds, and sedges, and is a massive structure, often standing a foot above the ground.

A familiar example of the order *Insectivora* and family *Turdidæ* is the Water-ouzel (*Cinclus aquaticus*) perhaps the most interesting of our mountain water-birds. Although sombre in colour, and not elegant in shape, yet its lively motions in the water render it a favourite with the naturalist. No non-web-footed bird succeeds better in swimming and diving. It has a pleasant, although simple, note. It makes its nest early in April; it is a large structure, made of grass or moss, having an opening at the side, somewhat of the shape and character of the nest of the Wood-warbler. It puts its nest under the shadow of banks, overhanging the water, often within reach of the spray of a waterfall. Mr. Hewitson says,—" My friend Mr. Benjamin Johnson has known of a nest of this bird for many years in succession, which was built upon the rafters in one of the salmon fish-locks upon the river Tyne. It has been repeatedly known to build its nest under the arch formed by a natural waterfall, or milldam, and within reach of the passing spray. The eggs are four or five in number, white when blown, but of a

WATER-OUZEL.

delicate pink when the yolk is yet in them. Once when in company with Mr. George Selby, in the beautiful grounds of Twizell, we came suddenly upon a nest full of young dippers, which, though scarcely able to fly, instantly scrambled into the water, down the stream of which they were hurried with such rapidity that I supposed it was impossible any of them could have escaped destruction. They did so, however, and landed safely far below." ("Oology," vol. i. p. 78.) Sir William Jardine, a practical naturalist, gives an excellent description of its habits in his native country. "The common water-crow, or pyet, as it is familiarly termed in Scotland, is a favourite with every one who resides near its haunts. The solitary and secluded nature of the streams it frequents, and their often wild character, render it a most fitting accompaniment, sufficient to break the solitude, but never obtruding on the calmness of the picture : one of those beautiful instances of nature's chaste compositions, where the life of the landscape combines to harmonize with all around, and here the effect is still more brought home by the simple and peculiar melody of its song. Its common locality in summer is rocky alpine or sub-alpine streams, and it seems indifferent whether the banks are thickly clothed with wood and natural brush, or are bare and barren. If civilization has encroached on their retreats, and machinery and mills have been in consequence erected, it accommodates itself to the event, loses its secluded habits, and seems to enjoy the bustle. It may often be seen perched on the inner spokes of the mill-wheel, singing its low melody, and we have known it to breed within the passage of the torrent which drove it. In such places they live in pairs, each having, as it were, a locality or limit within which they range, and select an appropriate situation for the nest. They sport about the banks of the stream, flying short distances, and during flight utter their single monotonous alarm or call-note. When about to alight, they drop or splash into the pools or stream, and almost never at once settle on the stones or rocks. They are one of our most pleasing songsters, though from the lowness of the note it is not often observed; but to the angler, who plies his rod at all hours and in the most sequestered scenes, it is a well-known and welcome strain. It may be at times heard during the whole year; but spring and the breeding season are the periods when it may be most easily and constantly enjoyed."

The birds being early breeders, this sign of the coming year is often heard in February, while the streams are still bound up in ice, and a clear shining morning at this early time will be sure to display some of those songsters, perched on a prominent stone or stick, or on the edge of a frozen pool, warbling their notes, just audible above the murmurs of the stream. In winter, when the higher streams become frozen and the cold intense, the "water-crow" removes to the banks of the larger and lower flowing rivers, or to the margins of some unfrozen lake. Here they find a more abundant supply of food, and their aquatic habits and manner of feeding are more easily observed. On every reach one or two may be now seen perched on some projecting stone or stick, or watching by the very edge of the ice, whence they drop at once on their prey, consisting at this time in great part of small fishes. They are most active in their motions during this occupation, and dive and return to their station with great rapidity. In milder weather, or when the rivers are less choked with ice, they swim and dive in the centre of the pools, and so expertly that we have mistaken and followed them for the Little Grebe. At this time, and I may say generally, aquatic insects, the larvæ of *phryganidæ*, or caddis-flies, and in some situations different species of fresh-water shells, form their chief food, which in summer again is varied by a greater choice of insects and aquatic larvæ. It has been during the continuance of a very severe frost only that we have seen this bird seize some small fishes in the manner above mentioned of diving from the edge of the ice; at the distance observed they appeared to be minnows, and were brought up held crossways in the bill. The ova of any kind of fish we have never detected in their stomachs or intestines, nor do we think they habitually at the proper season frequent the places where spawn would be deposited, and if they did, we would deem it almost impossible that they could reach it after it was impregnated and covered in the spawning-bed, which it is before the parent fish leaves the place of deposition. Neither have we any knowledge of the ova being sought after about the period when they begin to acquire vitality, and when they might become a much more easy prey. This, in fact, is the only time when any destruction could be accomplished. In the north of Scotland this little bird is persecuted for its supposed depredations, and we were astonished, before learning the reason, to find such suitable localities totally uninhabited by them. Here the provincial name of Kingfisher is given to them; a reward of sixpence is put upon their head, and in one Highland district we have the factor's authenticated report of five hundred and forty-eight having been destroyed in three years!" (Nat. Lib., vol. ii. pp. 70, 71.) The colour of the head and back of neck is umber; the wings, tail, and upper parts of the body blackish-gray; the throat, neck, and breast white; the belly chestnut, shading into a darker colour on the flanks, and to blackish-gray on the vent and under tail-covers; the legs and feet yellowish-gray. In the female the white and chestnut are less clear, and the gray edges of the feathers on the back are clouded. The young have the head and neck gray, the edges of the feathers being yellowish-white. The throat and breast are dusky yellow, or grayish-white, darker where the chestnut commences, but the edges of the feathers pale, and in the throat and breast the feathers are tipped with a narrow bar of blackish-gray, which gives a general dull-coloured appearance to the bird.

The only species of the family *Alcedinidæ* is the Kingfisher (*Alcedo ispida*), also the only example of the order *Fissirostres* which can be called a water-bird. From its brilliant colours and comparative scarcity it may be considered a feeble connecting link between the birds of northern and more brilliantly-coloured southern climes. It is found in every country in Europe. It is much more common in England than in Scotland. Its flight is very rapid. Its favourite haunts are small streams overhung by willows or other trees, from the boughs of which it can sit watching the movements of the minnows or sticklebacks, on which it principally feeds. It eats, likewise, water molluscs, leeches, and beetles. It excavates a hole in a bank or at a root of a tree for its nest. Mr. George Dawson Rowley has kindly furnished me with a nest and eggs of this kind, with the following note:—"Nest of the kingfisher with seven eggs, a good deal incubated; they were found in a chamber under an overhanging bank.

THE KINGFISHER.

The eggs were placed on fish-bones which, having been swallowed by the bird, were cast up. They were a full inch deep, and formed a deep drain, evidently placed there by design with the object of allowing moisture in the nest to escape; the whole ran upwards as usual. The chamber was very near the surface. The nest was taken at Backbrook, St. Neots, Hunts, April 20, 1861." I may add that Mr. Rowley was at one time a most intimate correspondent of mine, and furnished me with most valuable material for study and comparison in the way of nests, eggs, and young birds found in his native county of Huntingdon. I have had or seen about a dozen kingfishers' nests, and they all but one consisted of disarticulated fish-bones. The eggs are nearly round, of a beautiful pure white, with a lustre surpassing most white eggs, quite inimitable by art.

The Kingfisher has the mandible dark brown; the inside of the mouth, the base of the maxillæ, and the feet are orange-red colour. The head, wings, and upper parts of the body are olive-green; the edges of the feathers are bordered with bright verdigris-blue, which shows most brilliantly in the wings and back when the bird is flying. The quills are dark brown, edged with olive-green, as is the under part of the tail; the upper is brilliant blue. There is a line of dull blue from the maxilla to the back, and a spot of white on the neck. The chin and throat are yellowish-white, and the under parts orange-brown. In the female the colours are less decided and brilliant.

The Bearded Tit (family *Paridæ, Calamophilus biarmicus*) is a bird so characteristic of the fenny districts as to be deserving of a place in this work. It differs very materially in structure from the true tits, from which the genus was separated by Dr. Leach. Its habits are entirely aquatic. It delights to live among the reeds, and climbs among them almost after the manner of a quadruped. Its food consists mostly of *succinea*, *limnea*, and other small molluscs. The nest is generally formed of the common reed and sedge, and is of a cup shape. I possess one, however, where the form is that of a bottle. The eggs are white and shining, with small brown scratches, quite unlike those of any other British bird.

BEARDED TIT.

They were at one time pretty abundant on the banks of the Thames, but the destruction of the reed-beds has rendered them scarce. The neighbourhoods of Scoulton Mere and Yarmouth are their principal localities. It is one of the most beautiful of our British birds, and one which, from the draining of the fens, may be said to be becoming extinct in many localities where it was formerly plentiful. It is a perfectly harmless bird from an economic point of view, and very useful as a destroyer of the pests of our marshes.

The beak is orange, and the irides bright yellow; under and in front of the eyes is black, and the feathers at the sides of the cheeks hang down in a strip or moustache; hence its name. These can be raised when the bird is excited. The head, neck, and auriculars are blue-gray, shading into light pink on the chest. The back and lower parts are yellowish-brown, the two outer feathers gray, with a white edge; the secondaries dark brown, edged with orange-brown; the inner webs yellowish-white, with a longitudinal stripe. In the female the colours are less bright, the crown is dull brown; the tail has the inner webs of the three outer feathers brownish-black, and the under parts are pale brown or yellowish-white. The moustache is the same colour as the neighbouring feathers. The female is smaller than the male. Mr. Yarrell says the young have a black beak, and the wings and tail are patched with black, and the under surface fawn-colour. The moustache is a narrow black line. The full-grown bird is about 6 inches long.

The Marsh Titmouse (*Parus palustris*) has become rare in Scotland, but is still abundant in the fenny districts of England. It has so peculiar a note that it is not easily mistaken for any other bird. It frequents low, marshy districts, and generally builds in a pollard or old willow-tree. The nest is more carefully built than that of most water-birds. It is formed of grass and moss, and lined with the down of the willow. It has been known to make its nest in a rat's hole. There are seven or eight eggs, like those of the blue titmouse, but the spots are larger, and the shape of the eggs rounder. "It feeds on carrion," says Mr. Selby, "as well as on seeds, especially those of the sunflower and thistle." The head, nape of the neck, and throat are black. The head to the eyes has

THE MARSH TITMOUSE.

a distinct cowl or covering of a black colour. The upper parts and wings are yellowish-gray, the cheeks and breast nearly white; the lower parts are the same as the upper, but tinged with brown on the flanks.

The female does not differ materially from the male; the length is 4 inches 4 lines.

The next family, *Motacilla* or Wagtails, was originated by Latham, and seems a natural and justifiable order. The Gray Wagtail (*M. boarula*) is a partially migratory bird, which haunts the clear stream, and by its elegant form and sprightly motions enlivens the loneliest spot. They like to breed near a waterfall, either in some mountain or hilly district, or near the fall of a water-mill. The nest is usually on the ledge of a rock or bank, in a sheltered position. It is rudely made of stalks and roots, and lined with hair and fine grass. This bird feeds chiefly on insects and the smaller molluscs. It is generally distributed in the south of England in the summer, but is not very abundant. It usually has two broods during the season, commencing very early in the spring. It is common in the north of England, Scotland, and Ireland.

The upper parts are ash, the rump olive-yellow; above the eyes and on the sides of the neck there is a white band; the throat is deep black, the lower parts of a clear yellow; the wings and six intermediary quill-feathers of the tail are black, edged with white and olive; the three side quill-feathers of the tail white, the other two are black on the outer edges. The tail is $2\frac{1}{2}$ inches longer than the ends of the wings. The length of the male is 7 inches and 3 lines. The above is the spring plumage of the male. The female and the male, after the autumnal moult, have no black on the throat; it is reddish-white; the mark above the eyes is yellowish; the upper parts of the body of olive-gray, and the lower of a pale yellow. The eggs are very pointed at one end, very wide at the other, and are of a dirty ochreous white, with darker spots; they are from four to six in number.

The Gray-headed Wagtail (*M. flava*) arrives about the middle of April, and is often seen following the plough for the worms and grubs which are turned up. It builds in low, damp situations, seeking a dry tuft or spot in the marshy ground, or by the river-side, on which to place its nest. It sometimes makes its nest in old mole-holes in meadows; it lays usually six round eggs, of an olive-green with flesh-coloured spots.

The White Wagtail (*M. alba*) has the forehead, cheeks, and lower parts pure white; the back of the head, neck, throat, and breast, the feathers of the middle of the tail, and upper coverts of the tail deep black; the back and flanks ash; the coverts of the wings blackish, edged with white; the two outer feathers of the tail white. The length of the bird is 7 inches. In the female the white is less clear, and the black of the head does not cover so large a space, and the edges of the wings are grayish. The above describes the spring plumage of the sexes. There is a pure white variety (*M. albida*), also a variety sometimes with black wings and the remainder of the plumage of the ordinary appearance. But the winter plumage differs in that the throat and forepart are pure white, with a collar or band of jet-black on each side, bending up to the throat. The ash-coloured parts of the bird are paler than in the summer attire.

The young have the lower parts of a dirty white; across the breast there is a band of brown-ash; the remainder of the plumage is ash-colour. The young, hatched in spring, begin to take the adult plumage in autumn. The young of the second brood migrate before attaining the adult plumage, and even return in this condition; they are then (*M. cinerea*). They live on the borders of streams, or even in towns and villages, and on towers and steeples. They go as far as the Arctic zone. They build in meadows, on shelving, rocky places under bridges, or in towers or holes of trees. They lay sometimes six eggs, of a bluish-white, spotted with black. They feed on gnats, centipedes, woodlice, and insects and their larvæ.

The Yellow Wagtail (*M. Rayi*), builds its nests on the ground in fields or on the banks of rivers. The nest is made of dried grass, moss, or wool, and lined with hair and fine fibres; but Mr. Newton says the construction of the nest differs even in the same locality, sometimes being made of green moss and rabbits' down. The eggs are rather smaller than those of the Gray Wagtail, and much resemble those of the Gray-headed Wagtail; and even are so like the eggs of the Sedge-warbler that if mixed together Mr. Hewitson says it would be difficult to distinguish them. The head of this species is pale olive; it has a yellow line over the eye, and the chin and throat are yellow. The two outer feathers of the tail on each side are white, with a streak of black on the inner side; the other feathers are brownish-black. It very much resembles the Gray-headed Wagtail (*M. neglecta*), but the latter has a gray head; the line over the eye is white, and the chin white; the tail-feathers are a good deal alike, but the six central feathers of the latter are black. Ray's Wagtail (*M. campestris*) is a summer visitor to this country. Its nest is usually made in a hole in the ground, but is so carefully hid that it is seldom taken. These wagtails are not uncommon in Middlesex, but the Gray-headed Wag-

THE YELLOW WAGTAIL.

tail is very uncommon; it is mentioned as a British species by Mr. Yarrell, and quoted in Harting's "Birds of Middlesex." The Pied Wagtail (*M. Yarrelii*) builds in walls, bridges, and on the banks of rivers. Mr. Yarrell says it frequently builds in a woodstack or hayrick; and Mr. Jesse mentions a case in which it built in a brazier's workshop, and, notwithstanding the noise and peculiarity of the place it had chosen, the young were reared. The nest is made of roots and grass, lined with finer bits of the same and hairs. The eggs are often like those of the house-sparrow, and sometimes resemble those of Savi's Warbler. This common species is found almost everywhere in Britain throughout the year, but it migrates from north to south, according to season. All the wagtails frequent water, and feed much on water-insects and larvæ.

CHAPTER III.

A RAMBLE AMONG THE MARSH BIRDS OF SUSSEX.

On the 15th of May we started by train for Uckfield, that being the nearest station to the lodging where our friend Barnes had been staying for the last eight weeks. He met us about half a mile outside the village, where four roads cross. His three sturdy boys, between twelve and sixteen, came running up to us, and before we shook hands began exhibiting nests of eggs and an unfortunate bird with its legs tied together with a hair, vainly striving to get loose. We were not long in hearing from Barnes himself an account of his captures as we walked along together. Having breakfasted at the farmhouse where Barnes lodged, six of us started for a bird's-nesting day. It was just nine o'clock. "We'll visit the swamp first," said Barnes to me. It was raining slightly, but we tucked up our trousers, took each a basket or good-sized box, knives, mustard-spoons, string, a rope ladder invented by Barnes, and a first-rate old blind setter with an extraordinary gift for finding moor-hens' nests. We passed a deserted wren's nest in a rather high tree, which did not seem worth climbing for. "I have a list in my pocket," said he, "of what I have found, and ought to be ready for taking to-day.

I have beaten up the country well for the last month, and given rewards of from twopence to a shilling for

THE WREN'S NEST.

every nest I care for. I see them all myself. Most of my best nests I have reserved to show you *in situ* to-day. I allow no one to touch them; it is the safest way, and there is no mistake then. Bring me to the nest, boy, and you shall have your twopence, or whatever it is. Here we are close upon a nest —soft!"

Just then a Green Woodpecker (*Picus viridis*) flew from a large ash-tree and went laughing round it. The tree was sound to above the first fork, 9 feet from the ground. We saw no hole, but some chips of rotten wood thrown out round the tree convinced us that the woodpecker's nest must be near. Up went one of the boys and soon reached a large hole, down which he crammed his arm to the shoulder, but could find no bottom. Barnes next climbed up himself, but the hole was more than a yard deep, and suspecting that if there were any eggs the hen-bird was on them, he fastened a bit of cloth over it, and taking a mallet and chisel proceeded to cut a passage for his hand in the ash-tree in the quarter in which he thought the nest was. At last a small hole was made through the trunk, and Barnes put his hand in but could not reach the nest, which was deep in the heart of the tree, upon a sort of ledge. The hen-bird sat close at first, but as the light was let in through the hole she began to flutter, and at last made a dart at Barnes's hand, he caught her and drew her out; but respecting the words of Moses (Deuter. xxii. 6), he, after showing her to us, let her go. She flew all round, but at length settled on a tree not far off. The four eggs, much stained with the rotten interior of the tree, were laid upon chips of decayed wood, and about half a dozen feathers of the bird were laid under them. Barnes brought down the nest and four eggs, which he had extracted with the aid of a long spoon ; but before he had packed them up, his eldest boy shouted out, " I have found a Reed Warbler's nest !" (*Sylvia arundinacea*), and, lo! we saw

one suspended by three reeds, quite in the water, near a large willow overhanging a brook about fifteen yards off. It was a fine nest, and a good deal of sheep's wool was used in its construction, but it was rather beyond our reach. It was a pretty sight to see the nest waving to and fro in the reeds. The bird had flown, but Barnes and I crouched down in the long grass for about five minutes, soon to hear her shrill cry, and to see her jump among the reeds and climb the stem like a mouse to her nest. She looked into it and then flew away, probably catching a glimpse of us. I got on the willow-tree, and fixing my feet on a forked branch, I stretched myself to the utmost, although at some risk of falling into the stream. The breeze bent the reeds on one side towards me, and

REED BUNTING.

then I saw five green eggs laid out so prettily that I protest if I had not been anxious to study the materials of which it was composed I should have left it there to wave to and fro in the wind. It was shaped almost like a wine-glass, suspended among the reeds. I got a pair of scissors with very long handles, and taking a crooked walking-stick in one hand, and the scissors in the other, I held the nest towards me and cut the reeds one by one, and so got it in safety.

COMMON REED IN FLOWER.

We found here an empty nest of the Reed-Bunting, and saw several of the birds flying about. The flowering tops of the common Reed (*Arundo phrag-*

mites) were very abundant. The Fringed Waterlily (*Villarsia nymphæoides*), was common in this water, as in several of the brooks of Lewes.

FRINGED WATER-LILY.

As we entered the marsh we heard the Cuckoo (*Cuculus canorus*) for the first time that year, and were induced to examine every nest in the neighbourhood carefully, in hopes of finding an egg of this singular bird. We found two nests of the Sedge

Warbler (*Sylvia phragmitis*); one was empty, and a second contained an egg of the warbler with one of the cuckoo. The cuckoo's egg was quite warm, and the little nest was pressed down on one side, evidently by the feet of the heavy bird. The nest was on the ground amongst the grass and sedges. It was

THE CUCKOO.

pretty smooth inside, but less neat and elegant than the reed warbler's. In a little copse on the left-hand side of the road the boys pointed us out a nest of the Song Thrush (*Turdus musicus*), with five spotless eggs. It was placed in the fork of a laurel grown wild. These eggs we carefully preserved as abnormal.

In a little hedge of furze, so close that as we went through it we nearly tore our clothes, we disturbed a blackbird (*T. merula*) from its nest, with four young ones and an egg not hatched. I did not molest a goldfinch, which had her nest and four eggs nearly ready to chip, in the opposite fork of the same tree.

NEST OF THE GOLDFINCH.

We found a deserted linnet's nest (*Fringilla cannabina*) lined with the down of a peculiar thistle, a native of the pampas which ran riot in the garden of a cottage "to let" not far off. Barnes related how he had taken, the preceding May, a nest and eggs of Savi's Warbler in a fen in Cambridgeshire, and produced a photograph of the nest, which was entirely made of the common reed.

We turned into a very narrow lane about a yard wide. At one end was a stone wall, and the other end was nearly blocked up with briers. Long grass grew all over the path, and the brambles and clematis formed an arched roof. This was evidently a sanctum for the smaller birds. Barnes and I alone went in. We found a nest of the Wren (*Troglodytes Europæus*) just at the entrance, close to a stump entirely made of moss. In the ivy covering the stump, about 6 feet above our heads, we dug out the nest of a creeper (*Cesthia familiaris*), with five eggs hard set. But best of all was a willow wren's (*T. trochilus*) nest, the most beautiful I ever saw of this species, underneath an old coal-scuttle. The bird flew out, and we soon examined its six eggs, which were beautifully pink. A number of wrens were close to the stone wall, making a great clamour. We went very quickly, yet cautiously, towards them. In the hedge

NEST OF SAVI'S WARBLER.

sat a large common snake (*Tropidontus natrix*), in the act of swallowing a wren, whose nest was not far off in the withered leaves among the thorns. There was a good deal of earth used in it. It was firm, and the leaves of which it was made were dark chocolate-colour, being those of the beech. About half a mile off, in a high hawthorn, on an almost inaccessible branch, was a magpie's (*Corvus pica*) deserted nest. Some owls had been observed to hoot near the spot the evening before, but none could now be seen. We all shrank from a scramble through the thorns which obscured the nest, for a large branch of hawthorn hung over it. The rope ladder was fixed with some difficulty to one of the top branches. Barnes went up, and with his saw cut off the branch that overhung the nest, which gave him a view of a white moving mass through the thorns. He, with considerable difficulty and some expenditure of time, cut away the smaller branches which almost surrounded him, and was thus enabled to move in the direction of the nest; forcing away with his walking-stick the crown of thorns which covered it, he soon lowered down to us two young owls of the long-eared species (*Strix atus*) and two eggs quite fresh. The young birds were covered with down of a sort of dirty-white colour. We returned to the farmhouse, and had our lunch at two o'clock. In a very thick yew close to the house a pair of gold-crests (*Regulus flavicapillus*) had built, but the nest was so high up in the tree that there was great diffi-

GOLD-CREST AND NEST.

culty in getting it, and the landlord did not like his tree to be cut about. Not wishing to be balked, we with some trouble borrowed two ladders, each 18 feet long, which we steadied with ropes on either side of the tree, which was an enormous one. The second boy went up one of the ladders, and

NUTHATCH.

threw himself on the top of the tree, but it nearly *threw* him down by its elasticity. So we had to place a plank between the two ladders across the tree, astride which Barnes got, and with a pair of shears cut out the nest; the bird sitting so close that he could cover it with a butterfly-net. It was taken in the hand and let fly. There were six eggs. From Barnes's commanding situation he looked down upon a swift's (*Cypselus apus*) nest under the tiles of the house. This had been made the year before, but the entrance had been stopped up in accordance with some superstition of the mistress of the house. As I had never before seen a swift's nest, I got out of the garret window on the roof, and lifting two of the tiles, found it placed in a sort of hollow where a rafter had rotted away. There the skeleton of the bird lay, picked by ants, and two eggs which exploded at the touch. The nest was made of some glutinous material, but the rain had spoilt it.

The landlord's brother was a miller, who lived a few miles off. He was very fond of birds, and hearing that we were on an ornithological excursion, he drove us to his house, around which many species bred. He took us to a kingfisher's (*Alcedo ispida*) nest, the second of the year; but it was not formed of fish-bones, as is sometimes the case. It contained but one egg. The miller also showed us a nuthatch's (*Sitta Europæus*) nest in a fir-tree, which unfortunately was deserted; but he good-naturedly gave us four eggs taken from the same hole the year before. The nest was made of the inner bark of the silver birch, a rich treat to see, as it had been all stripped off by the bird. Our conductor having retired from business, had converted his mill-pond into a gathering-place for water-fowl. It extended to an eighth of an acre, and was surrounded by a thick copse. He had cut a canal round a portion of the copse, which made a secluded island, on which he had placed a bower, where we sat for some little time and watched the moorhens come ripple along the surface of the pond, and actually saw a coot catch and devour a frog. The wild ducks (*Anas boschas*) were sporting about, and we found a nest with seven eggs, which the miller did not want us to take, as it was the only one on the island, and he loved to see the young ducklings sport. A mute swan (*Cygnus olor*) glided along the canal. This, he said, he usually kept in another

NIGHTINGALE.

pond, as it created rather too much commotion among the smaller fowls. As he spoke it uttered a chucking sound, which I heard for the first time. As we rode home through the green lanes the nightingales (*Sylvia luscinia*) and the black-caps (*S. atricapilla*) sang a charming concluding chorus to a delightful day.

CHAPTER IV.

AQUATIC AND MARSH BIRDS. — HERONS, PLOVERS, STORKS, SANDPIPERS, SNIPES, RAILS, DIVERS, GREBES, GEESE, DUCKS.

THESE birds are less abundant in Britain than formerly, for if they swim on or wade in the water, their especial home in the breeding season is the marsh, which decreases as civilization increases. "The Bittern (*A. stellaris*, Linn.), in the good old times, when cornfields were marshes, and market-gardens yielded only the bulrush and the sedge, often uttered its wild, peculiar note in the Eastern Counties; but now the bird is rare—a chance visitor from Holland or the north of Europe. When 'the bittern possessed the land' it was laid waste, for this bird cannot exist in the midst of civilization; yet when we see it in the midst of woe, and hear its wail among ruins, we contrast its cry with the sweet warblers that enliven the pleasant garden. The Bittern's melancholy cry does not cheer us, but it tells of the rise and fall of cities and empires, and of the wail of those led into slavery, and teaches us a lesson that none of the song-birds can do, although their melody may carry our ears captive the whole summer, for they speak in simple accents of bright days, present or to come."[1]

---

[1] 'Book of Nature and Book of Man,' page 242.

THE BITTERN.

The Bittern is 2 feet 5 inches long, from the tail to the tip of the beak, and is of a yellowish colour,

THE BITTERN.

very prettily marked with a zigzag pattern of brown. It is unlike any other European bird of its family in colour. Its nest, which has been once or twice taken in Norfolk in living memory, is a loose structure, formed of aquatic plants, and contains from three to five eggs of a very dull greenish or yellowish-white, faintly washed with ochre. The Bittern is not very uncommon as a winter visitor to Britain, more especially in stormy weather, for the influence of winds diverts it from its natural course between Arctic and Southern Europe, its stronghold being Hungary and the marshes of Lithuania.

Among the birds of the family which are now very scarce, but which in former times inhabited our marshy districts, may be mentioned the little Egret, *Ardea garzetta*, an exquisite white species, 1 foot 10 inches from beak to tail. This bird is decorated, in the breeding season, by long plumes, or *aigrettes*, proceeding from the back of the head, and by very remarkable hair-like feathers about the wings and breast; the latter are the heron's plumes, which were so eagerly sought in ancient and modern times for decorating the human head, and which are now sometimes set with sprays of diamonds by the most fashionable jewellers. This exquisite bird, from the efforts to obtain it in the breeding season, as well as by the drainage of the marshes, which has destroyed the sanctuaries of its nest, has become almost extinct in England, for it is only after very long intervals that specimens are obtained. The same may be said of the much larger Great White Heron (*Ardea egretta*), which resembles it in colour. This bird is one of our rarest visitors, and we have no record that it ever bred in Britain. Its stronghold is in the north of Africa, Syria, and Hungary.

The lovely Purple Heron (*Ardea purpura*), a bird of most brilliant purple and brown plumage, is less rarely seen in England, but still it is a bird accidental in its visits. The same remarks apply to the Night Heron (*Ardea nycticorax*), a bird of very beautiful gray plumage, and of a shorter, more compact build of body than those we have described. The Squacco Heron (*Ardea ralloides*) is a bird of most beautiful delicate brown and gray plumage, and carries from its head long plumes; it is a small species, 16 inches in length.

The Buff-backed Heron (*Ardea bubulcus*) is a whitish bird washed with buff, about the size of the little egret, and about equally rare. The little bittern, which is much the smallest of the heron family, is only 13 inches long, and does not exceed in weight a young pigeon; it lays white eggs, whereas all the other herons of Europe lay blue or greenish eggs. It is a pretty little bird, quite unlike any other European species. The Common Heron (*Ardea cinerea*) is a bird with which most country folks are familiar. Its long neck and bright gray plumage render it a conspicuous object on the wing. It is strictly preserved in many heronries. It is of a bluish ash-colour, with long black plumes on the back of the head, and feathers of a shining white hanging from the lower part of the neck, and those of the scapularies are of a silvery gray; the forehead, neck, middle of the belly, edge of the wings and thighs of pure white, the sides of the breast and flanks of a deep black, in front of the neck large longitudinal spots of black and ash-colour; the back and wings of bluish gray, the beak of deep yellow, the iris yellow. The skin round the eyes is without feathers and is of a bluish-purple, the feet brown, but deep red towards the feathery part of

THE NIGHT HERON.

the leg. The length is three feet or more. Such is the colouring of the male and female after three years of age. But the young have no crest, or the feathers are extremely short, and the long neck-feathers are wanting. The throat is white, the neck of a clear ash, with deeper spots, the back and wings of ash-blue mixed with white and brown, and marked with long spots. The upper mandible of the beak is blackish brown with yellow spots, the lower mandible is yellow, the iris is yellow, and greenish yellow surrounds the eyes. The feet are ash-gray, but yellow towards the feathers. There is a perfectly white variety of this bird, which is very rare, but easily distinguishable from the great white heron by the nakedness of the legs.

In former times the heron was protected by law, as it was much sought after in hawking. In those days it was very abundant, and is still not rare; its striking appearance makes it easily recognisable by the most casual observer of bird-life. Its general outline at some little distance makes it appear more grand and formidable than it really is. It stands higher than the golden eagle, though the expanse of its wings is much less. Its length is about 40 inches, and its breadth about 5½ feet. It has a peculiar way of doubling up its long neck when flying and throwing back its feet, which makes it appear very long, and distinguishes it when flying from the goose, for which, if it flew with the neck stretched out, it might be mistaken. It flies high, though less so than the birds of prey. Seen nearly vertically above the observer, its flight seems rather graceful, but the reverse is the case if the beholder looks down on the bird from an elevation. Were we to estimate its weight in proportion to cubic size, in order to give to the heron the steady and powerful flight of a raptorial bird of the same dimensions, it ought to weigh much heavier than it does, but its weight is only about three pounds. It is, indeed, the lightest of British birds in proportion to its dimensions. The Barn Owl (*Strix flammea*) is the lightest of the owls, considering the size of its wings; compared with the size and weight of the heron, however, it is proportionately much heavier. The soft feathers of the owl, while adding little weight, from their breadth and shape take hold of the air in flight and help almost noiselessly to propel the bird. The feathers of the body of the heron, however, lie very close to the body, and do not much assist in flight. This peculiarity causes the water to run off the bird and adapts it to its aquatic life. The bones of the wing are hollow, and in descent the exertion of much force is needed, and the body is thereby made to jerk in an awkward manner. Less force is required in rising from the ground or whenever it is perched, but still this peculiarity of form and the closeness of the feathers give an ungraceful up-and-down motion to the flight. Yet this habit is useful, as the throwing of the neck backward makes the centre of gravity fall on the thighs as the bird raises itself, and enables it to take wing at the first stroke, even when on the ground or when standing in shallow water. The heron has often to fly a great distance to get its food for itself and its young, therefore it has to soar very high, in order to see the fishing-ground it requires: this is usually in the early morning. It is a conspicuous object for hawks and falcons, but the eagles of Britain being now so rare, it seldom runs any risk from them. The bill of the heron, nearly 6 inches in length, is so powerful that on the ground the bird would probably be a match for the hawks. The herons are gregarious in their nest-building; they select trees near a stream or piece of water, and they like to build in the same locality year after year. Their nests are large and flat, composed of sticks, and covered with rushes or long, dried grass; they lay from four to six eggs, about the size of those of a duck. They are apt to settle on the tops of the trees in which their nests are during the day, and now and then stretch their long necks and spread their wings like the indolent among mankind, who yawn and stretch themselves through lassitude or idleness. This is the habit of the birds in fine weather; when the sky threatens rain and the storm lours, they are more active, watching after a flood of rain for the fish which may have been washed down by the torrent. Their bills are sharp, and are furnished with rows of barbs or small teeth, which enable them to hold fast any prey they may have caught, even if slippery as an eel. They are so skilful as anglers that in one hour they will catch more small fry than could be done in a much longer space by any human angler of ordinary skill. They injure many fish, which they do not catch, by the sharpness and power of their bills. It is commonly reported that small eels pass through the heron undigested, and that it may thus swallow the same fish more than once; this is not however the case, but unless the bird kills the eel at once by laying hold of its head or gills, it carries the eel to the ground, and setting its foot upon the wriggling fish, soon squeezes its different parts by its powerful beak, and swallows down what to the bird is a delicious meal. The male feeds its mate, as well as helps to feed the young during incubation. The heron, though it roosts on trees, is yet like a web-footed bird in its habits. Yarrell mentions a curious instance of a heron, which having captured an eel, was, before it could swallow it, suffocated by the fish twisting, in its contortions, round its neck like a snake.

The Plover family, or *Charàdriidæ*, are important

THE BLACK-WINGED STILT.

marsh birds; the Stone Plover, which breeds on moors, being an exception.

The Black-winged Stilt (*Himantopus melanopterus*) has, perhaps, longer legs, proportionately, than any other bird. It is smaller than the lapwing, weighing only five ounces; it is only an occasional visitor to England.

The Golden Plover (*Charadrius pluvialis*) is a moorland bird; and the Dottrell (*C. morinellus*) is an alpine bird which appears in the British Isles in the breeding season, but they are both addicted to marshes in winter.

GOLDEN PLOVER.

The Ringed Plover (*C. piaticula*) is a bird of the sea-coast, breeding there. The same may be said of the Kentish Plover (*C. cantianus*), Sanderling (*Tringa arenaria*), Turnstone (*Strepsilas collaris*), and Little Ringed Plover (*C. minor*). These birds are occasionally marsh birds.

The Lapwing (*Vanellus cristatus*) is a bird that lives rather on the moor than the marsh.

The Oyster-catcher, of the family *Hæmátophidæ* (*Hæmatopus astralegus*), weighs about a pound, and is about 18 inches in length. It is a bird of beautiful plumage, the prevailing colours being black and white, which form a pleasing contrast to the bright red bill and feet. This bird measures nearly $1\frac{1}{2}$ feet long, and in expanse of wing more than $2\frac{1}{2}$ feet. It is often met with on all parts of the coast. The bill has a length of about 3 inches, with nasal grooves an inch and a half long, and nostrils of longitudinal form. Its long legs are bare for about an inch above the tarsal joint; the toes are partially webbed. It can walk equally well on rough or slippery ground, for the under part of the foot has small protuberances. The wings are long and pointed, and the tail is long and square at the extremity. The plumage is very close, and difficult to wet. This bird can swim, though it is not an habitual swimmer. The sea-shore is the haunt of the Oyster-catcher, but it feeds on fresh-water mollusca, and land-slugs and worms. The female sits closely at night and when it rains, and the male is always near, ready to scream and fly off when danger comes. The two then continue to hover about, making a noise till the danger is over. On warm days she leaves the nest to feed. The incubation occupies three weeks. The young, at first, walk with difficulty, but they are soon able to run and fly.

The Oyster-catcher lays from two to four eggs on the ground; it is partial to heaps of broken shells or pottery; it makes or takes advantage of a slight depression in the ground in which to place its eggs, which are of two or three shades of stone-colour, sometimes inclining to green, the spots being ash or nearly black, and differing greatly in form, size, and disposition; one variety has a few red spots. When there are four eggs,

THE OYSTER-CATCHER.

the bird places them in the form of a cross. There are few counties possessing a sea-coast where the Oyster-catcher is not found; but its favourite localities are Norfolk, the Fern Islands, the Isle of Man, and the Isles of Scilly and Lundy, and the west coast of

Ireland. Its eggs are excellent eating, and are laid from April to July.

Of the genus *Ciconia*, the White Stork (*Ciconia*

THE WHITE STORK.

alba), and the Black Stork (*C. nigra*), and the Crane (*Grus cinerea*) are the chief examples, but they rarely visit Great Britain. The latter was protected by statute in its breeding season in Norfolk, but as it has not bred there for 200 years, this law may well be looked upon as obsolete.

The Common Crane (*Grus cinerea*) of the family *Gruidæ*, visits Britain occasionally in winter, as do the storks. They pause to feed in rivers or on their banks, and are destroyed by numerous sportsmen. The same remark may apply to the Glossy Ibis (*Ibis falchionellus*), a bird of rich purple and greenish plumage, but which has never yet been known to breed in Britain. It is blown to our shores by the gales of winter.

The Curlew (*Numenius arcuata*, family *Scolopacidæ*) and Whimbril (*N. phæpus*) are moor and mountain birds, but yet breed in the neighbourhood of our lakes occasionally; the Sandpipers in Britain are more strictly aquatic birds; of these the most common is the Redshank (*Totanus calidris*, genus *Totanus*). This is indigenous, and not uncommon on boggy grounds, to which the birds resort in April and May. It makes a rude nest, by gathering together dry grass beneath a tuft. It lays four eggs, which vary in ground colours from greenish-yellow to olive-brown; the spots are very dark reddish-brown and liver-coloured; they are sometimes found in a zone, at other times greatly gathered together at the large end. They are $1\frac{1}{16}$ inches to $1\frac{4}{16}$ inches long, by $1\frac{7}{32}$ inches to $1\frac{9}{32}$ inches broad.

The Redshank is not gregarious in winter, but leads a solitary life along the coast. The feet and part of the bill are yellowish-red. The bird is 11 inches long, the expanse of wings is 21 inches, and the weight five ounces. Like most of the water-birds, there is a conspicuous change between the summer and winter plumage. In summer the head and back of the neck are brown-ash, with dusky streaks in the length of the feathers, and a white streak over the eye. The back and scapulars are dusky, with dull gray

THE WHIMBRIL.

spots; the coverts ash-colour, with spots of brown and white; the rump white, marked with small spots and bars of dusky brown; the tail is barred with black and white. In winter the plumage of the back changes to ash-brown with dusky streaks, that of the breast to pale greenish-white and slender brown streaks. The young have the upper plumage brownish, the plumage on the breast ash-colour, with pale brownish streaks, and the tail-feathers have reddish-brown tips. The Spotted Redshank (*Totanus fuscus*), which does not breed in Britain, is a bird of the river banks, but a winter visitor; its breeding-ground is in Norway and Lapland.

The Greenshank (*Totanus glottis*) is a bird of similar habits, and visits this country. Dr. McGillivray found it breeding on the lakes in Unst, Harris, and Lewis. In the summer season " it is easily discovered, for when, perhaps, more than a quarter of a mile distant it rises into the air with clamorous cries, alarming all the birds in its neighbourhood; flies round the place of its nest, now wheeling off to a distance, again advancing, and at intervals alighting by the edge of the lake, whence it continues its cries, vibrating its body all the while. The nest found in the Island of Harris was at a considerable distance from a small lake, and consisted of a few fragments of heath and some blades of grass placed in a shallow cavity, scraped in the turf in an exposed place, that is, on a slight eminence, covered chiefly with moss, lichens, and some carices (sedges) and short heath. The nest, in fact, resembled those of the golden plover, lapwing, and curlew." " In 1836 Mr. Selby found this species breeding in various parts of Sutherland, generally in some swampy marsh or by the margin of some of its numerous lochs. It is very wild and wary, except when it has tender young, at which time, when first disturbed, it sometimes approaches pretty near, making a rapid stoop, like the redshank, at the head of the intruder. If fired at and missed, which is frequently the case, even with a good marksman, as the stoop is made with remarkable rapidity, it seldom, at least, for that day, ventures again within range. A pair, which had their nest in a marsh near Tongue, after being once fired at, could not again be approached; but we obtained one of the young, apparently about a fortnight old, by means of a water-dog. Another pair were shot near Scourie, by the margin of a small loch, where, from their violent outcries and alarm, they evidently had their nest or young, though we were unable to find either."

The eggs of the greenshank are extremely beautiful, and usually of a pear shape; the colour is greenish, ochrous or cream-coloured in ground, with spots of lilac and purple-brown; they are considerably larger than the redshank's. The greenshank is 14 inches in length, and nearly 2 feet in expanse of wings; although a bird of plain plumage, it has an elegant and graceful form. The bill is $2\frac{1}{2}$ inches long, slender and black; the nape and sides of the eye are ash, the back ash, glossed with bronze-brown on the centres of the feathers, scapular coverts the same, quills dusky with white spots on the inner webs, belly, upper and under tail-coverts, rump and tail white, the latter crossed by irregular lines of a dusky hue. The legs are dark green; hence the name greenshank.

The Common Sandpiper is an abundant bird on the estuaries of our rivers and on the shores of our lakes, as well as on the sea-coast; hence it may be advisable to describe it in some detail. It is partly web-footed, but can run rapidly. Although not an

THE SANDPIPER.

habitual swimmer or diver, it can perform both these operations when alarmed, or occasion seems to require it. The wings are longer than those of other species of this genus, and the tail being fan-shaped gives it a peculiar motion when flying, or when on the

COMMON SANDPIPER.

ground. The Sandpiper (*T. stagnatilis*) migrates to various parts of Britain, and to some of the western isles of Scotland, but does not go very far north. They are extremely active, and have a shrill and somewhat plaintive cry. The nest is not elaborately formed, but is generally made of dry leaves, moss, or roots, and is placed in a sheltered situation near the ground. Sometimes the eggs are simply laid in a cavity among sand or gravel. Here the similarity of colour causes them to be not easily seen. They are of a deep cream, stone, or pale blue colour, spotted with ash and madder brown; they vary less than those of most waders, and are found in May and June. It is widely distributed, but the eggs are less abundant than those of the snipe

PURPLE SANDPIPER.

(genus *Scolopax, S. gallinago*) and dunlin; they are from $1\frac{6}{16}$ inches to $1\frac{2}{16}$ long, by $\frac{15}{16}$ inch broad.

The Dunlin (genus *Tringa, Tringa variabilis*), breeds largely on the shore of the inland lakes of Scotland and the north of England. The name *variabilis* has been given to it on account of its extreme diversity of plumage. Its winter dress is much paler than its summer dress. The head and all the upper part of the back are ash-gray tinged with brown, the shafts of the feathers inclining to black. The coverts are of a dingy brown with gray margins, and the ends of the large feathers whitish. The rump and upper tail-coverts are dusky-brown, the edges being lighter. The middle and longer tail-feathers are brown, the shorter gray with whitish shafts. There

CURLEW SANDPIPER.

is a brown line from the eye to the corner of the mouth, and above the eye a line of white; the cheeks are white and brown, the chin and throat white, the upper breast gray and the lower white. In this plumage the bird is called by some writers the "purre." In summer the head, back of the neck and scapulars become black, the feathers edged with reddish-brown, the lower part of the back becomes dark, the chin and flanks remain white, the cheeks, front of the neck and breast become black with white edges to the feathers, and the rest of the lower part of the body becomes black. In this state it is called the Dunlin. The young have intermediate colours. The purre or dunlin is between 7 and 8 inches in length, and about 15 in expanse of wing; the bill is of a black colour, and is as long as the head; the legs are a dull

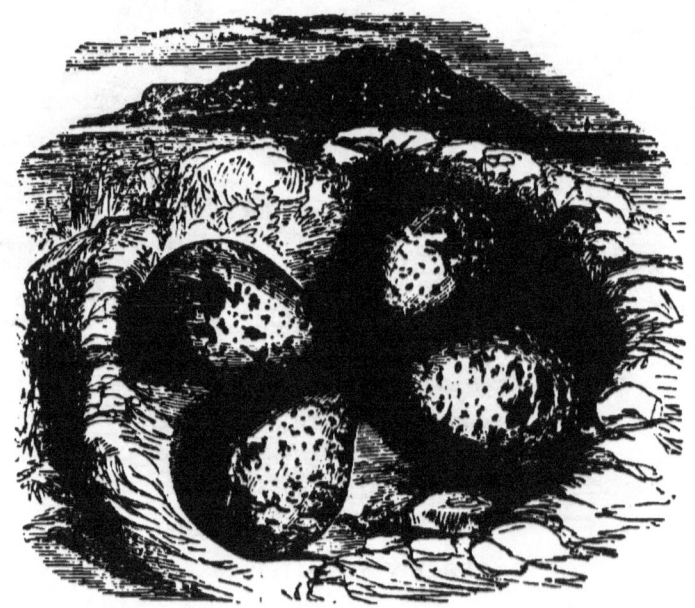

NEST OF THE DUNLIN.

greenish hue. It is abundant in winter on moist sandy shores, where it picks up its food as the tide goes out. It breeds freely in the north of England and Scotland. Its nest is simply a hollow scraped in the ground, lined with a little grass or dry twigs. When fat, these birds are considered good for the

THE KNOT.

table. Its eggs, which vary much in shape, are from $1\frac{5}{16}$ inches to $1\frac{7}{16}$ inches by $1\frac{4}{16}$ inch to 1 inch wide, and vary much in colour; one variety has the ground pale blue-green, with spots of purple, umber, red-brown, and purple-ash; another has a cream-coloured ground, with spots of raw umber and purple-ash; others have the ground yellowish with the same kind of spots, or with the spots reddish-brown; they vary much in form and arrangement.

KNOT.

The Knot (*Tringa Canutus*) comes to the fenny districts from the Arctic regions, where it breeds in the month of August. It was said to have been a favourite article of food with our Danish king Canute, of which the English name is a corruption. It is larger than the dunlin or sandpiper. It is more than 10 inches long, about 19 inches in expanse of wing, and weighs only four or five ounces. The bill is not so long as that of the purple sandpiper, the hind toe turns inward, its feet are well fitted to walk on soft ground, and in winter it frequents the sands and fenny districts. It changes its colours much according to season, so that it is known by a variety of names. Like other marsh birds it is much less plentiful in England than formerly, but in Lincolnshire and Cambridgeshire there are still considerable flocks of them to be seen. On the fens they collect in August. Dr. Fleming mentions having shot a specimen in June, 1808, in one of the Orkneys. They are to be met with on the sandy coast of Aberdeenshire, and they breed frequently, according to Mudie, in various parts of the country; they are shy, and it is difficult to find their breeding-places. But later observers have not confirmed their breeding in these places. It may have ceased to breed in these islands; its true breeding-place is the shore of the Polar seas, where Captain Nares's expedition found the young birds, but were too late for the eggs.

One of the best known of our marsh birds is the Ruff (*Machetes pugnax*, genus *Machetes*); it is a polygamous bird, and the males are very quarrelsome, and much larger than the females. Their most distinguishing characteristic is a large ruff of feathers over the breast and neck, and a tuft of feathers behind each eye. The ruff varies so much in plumage that it is almost impossible to describe its colour. The male is between 11 and 12 inches long, the female a third less.

THE RUFF.

This bird has declined greatly in numbers since the draining of the marshes in our eastern counties. They arrive in England in spring, the males appearing first, and do not quarrel among themselves until their jealousy is excited by the coming of the females. "They assemble on a rising knoll," says Mudie, "and battle for the surrounding spot and the lady—not in bands under leaders as some have alleged, but each single-

RUFF.

handed, or rather single-billed, for himself. The contest often lasts for several days, or is renewed on several mornings; but whether the victor of each day leads off a female at the close of the warfare of the same, or whether the same female occasions a contest of several days' operation, has not been determined. The battles of the hill continue, however, till all are mated to their desert, by which time the hill itself is often trodden like a pathway. The nests are rudely formed of withered grass, in the hassocks or tufts, which are separated from each other by sludgy or miry places; the eggs are four, olive-brown, spotted with darker brown; and the young are hatched about the middle or toward the end of June. During the whole of that period the males 'hill' in the morning; and they stand accused of some Don Giovanni-ism, but as the period of the young breaking the shell approaches, they 'hill' in fewer and fewer numbers, combat less energetically, and at that time cease from their combats altogether. It does not appear that the males take any share in the building of the nest, the incubation, or the feeding of the female while sitting, nor have they been seen tending the brood after. They are but little seen during the moulting month; and when they again make their appearance they are without their insignia of war, and withal very peaceable birds, and harmonious with each other." The eggs are laid in May, and are from 1 inch $\frac{1}{16}$, to 1 inch $\frac{1 \cdot 4}{16}$ long, by 1 inch $\frac{3 \cdot 1}{16}$, to 1 inch $\frac{5}{16}$ broad. They are subject to three principal varieties; one has the ground olive-brown, with darker spots of the same colour intermixed with those of purplish-ash; a second has the ground pale green with spots of burnt umber and purplish-ash; a third has the ground stone-colour, with spots of dark reddish-brown; the spots vary in size and shape. The ruff, now scarce, was once abundant in the fenny districts.

The Woodcock (*Scolopax rusticola*, genus *Scolopax*), at once a favourite with sportsmen and epicures, is mostly a winter visitor to the British Isles. Instances are known of its breeding in this country, and then usually in the vicinity of lakes or small pools. It places its eggs on the

WOODCOCK.

dead leaves of ferns or other plants; they are four in number, and vary considerably in colouring and size. One variety has a ground cream-white with spots and blotches of olive-brown and purplish-ash; a second has the ground pale olive-brown with darker spots of the same colour; in a third the ground is cream-colour with purplish-ash, intermixed with spots of pale yellow ochre. The eggs are from $1\frac{8\frac{1}{2}}{10}$ inches, to $1\frac{3}{10}$ inches long, by $1\frac{4\frac{1}{2}}{10}$ inches broad. In one case an egg was found almost without dark spots.

COMMON SNIPE.

The Snipe (*Scolopax gallinago*) is, including the bill, about 12 inches long, and the expanse of wing is about 14 inches; the weight of the bird about four ounces. The colour of the bill changes much after death, losing its smoothness and becoming of a dull colour, whereas in life the base is red, shading into yellow and brown towards the tip. The head is dark-brown with a light yellow stripe down the middle, a light stripe over the eye, and a brown line or spots down the side of the neck. The back is shiny black, and the wings dusky; the scapulars being marked with yellow. The tail-feathers are barred with black, and the coverts of reddish-brown. The front of the neck is yellowish-white with brown marks, the belly is white. The feet are greenish-ash. These birds when flying to a distance rise high in the air, so as to be out of sight, when even their cry can be heard. They are numerous in marshy places in the winter season. They have enemies in the hawks and marsh harrier, but their cunning manner of flight generally enables them to get out of the way in time. They quit the low places when the weather gets mild, migrating to exposed and cold situations. Many come from more northern countries in the winter, while some content themselves with migrations from one part of Britain to another. They are very partial to the boggy districts of Ireland, the Yorkshire Wolds, and Highlands of Scotland. The pairing cry of the male is a peculiar bleating sound, uttered in flight, and sometimes will continue all day, but when he has found his mate he is silent, except in the evening. The peculiar note continues during incubation. The nest is concealed in thick grass, or low growing plants, and is lined with dried fibres or leaves. The eggs are pale green-gray with brown blotches; the female lays usually four; the colour is sometimes considerably darker. The downy birds leave the nest as soon as hatched, and soon gain their feathers. The first plumage is darker than that of the parents. The bills do not grow to the proper length the first year. The birds sit so close that one may be near without disturbing or perceiving them, and when they are moved by the approach of danger, they quietly glide through the marsh-plants, scarcely showing themselves till they get quite off. They are

GREAT SNIPE.

not solitary like the bittern, though somewhat shy. They often have a difficulty in finding their proper food, but still their numbers do not much decrease.

The snipe breeds in small numbers in the fens of Norfolk and Suffolk, in Yorkshire, and some parts of the Highlands. It lays four eggs, which are from $1\frac{1}{2}$

to $1\frac{3}{16}$ inches long, and from $1\frac{2}{16}$ inches to $1\frac{3}{16}$ inches broad. They vary much in colour; one variety has the ground olive-brown, with darker spots of the same colour intermixed, with a few of liver colour; another has the ground stone-colour spotted with lines and vandyke brown; a third has a pale blue ground with spots. Sometimes they are very richly marked. They are found from April to July.

JACK SNIPE.

The Jack Snipe (*S. gallinula*) is a winter visitor to Britain; it is about half the size of the common snipe; its feathers, purple, bronze and green, have a metallic lustre. It arrives in Britain in September, and remains till February or March, when it goes north to Lapland and other Arctic regions for breeding. It is said to have bred in Great Britain, but this is doubted by most naturalists. It is fond of nestling on the low grounds, and it is therefore thought by some that it may hide itself in out-of-the-way marshy spots, where it would be difficult to find in the breeding season.

All the snipes quit the shores to breed inland. They fly chiefly by night, which adds to the difficulty of ascertaining the locality of shy and migratory birds. They like bogs which have here and there tufts of grass and sedge, forming little islands in the sea of swamp and mud, and where they are safe from the boat as well as from the pedestrian. The unfrequented cold, damp country, where plants decay and there is progress towards desolation, is unfavourable for species of this kind, as in time it affords neither food nor shelter. The eggs of the Jack Snipe are similar in markings, but a little smaller than the common snipe.

The Black-tailed Godwit (*Limosa melanura*, genus *Limosa*) is one of those birds which are excellent for the table, and which our ancestors formerly kept in cages and fattened for their feasts. The bill curves slightly upwards, like the avocet. In its breeding plumage the head is reddish-brown, streaked with black, and dusky; the lower part of the neck behind, the back and scapulars, are black barred with brown. A dull white streak passes over the eye, below which the cheeks, neck, and breast are pale reddish-brown. In winter the brown changes to gray, and the black to brown. It varies much in size, the average being about 17 inches in length, by about 2 feet in expanse of wing. Godwits are very shy birds, hiding during the day among the plants of the fens, and coming out to feed after dark, a fact which causes them to appear more rare than they really are.

The black-tailed godwit forms its nest in the thickest recesses of the reeds and rushes of dry grass and sedges. It lays four eggs, which are from 2 inches to $2\frac{6}{10}$ inches long, by $1\frac{1}{10}$ inch to $1\frac{7\frac{1}{2}}{10}$ inch broad. They vary from dark stone-colour to olive-green and olive-brown, and are variously spotted with olive-green, olive-brown, and ash-brown. This bird is much less abundant than formerly.

The Bar-tailed Godwit (*Limosa rufa*) is a winter visitor to our marshes, but individuals have been shot in the summer plumage, which renders it possible that it may occasionally breed in Britain, although its true breeding-place is within the Arctic circle.

The Avocet (*Recurvirostra avocetta*, genus *Recurvirostra*) is a most graceful bird, distinguished by its pointed bill, which curves upwards. It is about 18 inches long, and the expanse of its wings is 2 feet 6 inches. Its feathers are particularly beautiful, and are black and white; the legs are very long and the feet webbed; it is quite unlike any other British bird. It was formerly not unfrequent in the fenny districts, but now rarely occurs.

The avocet's habits, like its structure, are peculiar. According to Mudie, the little runs or watercourses which cross the loose sand or sludge, and which always contain a considerable quantity of spawn, larvæ, or other animal matters, according to the time of year, are the places which it frequents. It can swim, as indeed all birds that have close plumage on the under part can do more or less; but it perhaps does not swim voluntarily in any instance, and it never swims when it is feeding. It is not adapted for that, as the action of both the body and the bill requires a fulcrum of something more stable than water. Swimming in still water, the bird could not scoop, as the stroke of the bill would merely drive the body backwards; and as it feeds against the stream, its moving would be like that of a man attempting to force a boat against the stream by placing his pole upwards, and by that means adding his own exertions to the downward force of the current. The avocet wades up the shallow stream, and only that its strokes are equally effective right and left, its action is not unlike that of a mower. Its legs are long and placed far asunder, and it proceeds by long and slow strides. Suppose the foot advanced on one side and planted, and the other foot in the rear to the full extent of its stride, the axis of its body will in that position be across the run, with the head toward the side of the rear foot, and the tail to that of the advanced one, both feet being nearly in the line of the centre of the run; and if we suppose the left foot to be the one in advance, the bill will be over the right side of the run. The bird then bends its head a little to the neck and downwards, and immediately advancing the right foot, it swings the body upon the left as a pivot, the bill scooping a traverse curve and impelled by the swing of the body. As soon as the right foot is planted, or rather contemporaneously with the planting of it, the bird elevates its bill, in order that whatever food has been scooped up by the bill may be conveyed to the mouth; and that part of the process is very soon over, as the curve of the bill is not a portion of a circle, but of what the geometers call the "curve of quickest descent." The bill is immediately lowered with the point toward the right, and the advance of the left foot and the swing of the body upon the right make another sweep in the opposite direction. In this way the bird advances up the run, scooping alternately left and right with ease, with effect, and even with a grace almost unparalleled in the action of birds. It is indeed one of the most beautiful instances of animal mechanics that can possibly be imagined, and the motions are so performed that they can be all seen. Avocets are restless and lively in their manners, more sportive than most of the other fen birds. They have not the hiding disposition of the snipes, nor the demureness of the godwits; in some of their habits they more resemble the lapwing, especially in the finesse shown by the female to entice strangers away from her eggs or young. She meets the traveller, and flies round him in rapid circles, screaming "*Quheet, quheet,*" but aspirated in a manner that cannot be expressed by letters. She also runs and limps and drops one wing occasionally, as if it were broken; but in her evolutions upon the wing she does not give those twitches in turning which are so striking in the lapwing.

The *Rallidæ* or Crakes are generally summer visitors to the British Isles, and mostly breed in woods in the neighbourhood of marshes or rivers. The

Corn Crake (*Grex pratensis*) arrives in April, and greatly frequents the corn-fields, which afford, among the lower forms of animal life which feed on the corn, its favourite food. The male utters the cry, "*Crecq, crecq!*" until the young are hatched. The nest is hid in a thick covert; it is of rude construction, only a little dry grass and moss carelessly placed. If the long grass or brushwood is cut down near the nest, the birds forsake it, even if it is not molested, though the female sits so close that she might be killed by a scythe before she is observed. The number of eggs laid varies from eight to sixteen. The ground-colour is reddish-white, spotted with ash and rusty brown. The young birds are able to run as soon as hatched, but they cannot fly for some time. Except when about to migrate, they are not much given to flying; this is in the month of September. They have a habit of moving in a stealthy and almost noiseless manner through the long grass and bushes. They are very noisy in their cries, except when they are anxious to hide, but cease their cries for some time before migration. They feed at night, early in the morning, or late in the evening. Inharmonious as their cry is, yet in some lonely districts it is cheerful to hear them in the summer evening, especially as to the traveller it indicates the neighbourhood of cultivation and the quiet dwellings of some retired human inhabitants.

The Spotted Crake (*Grex porzana*), also a summer visitor, is a smaller bird. It is of a greenish olive-brown, with spots and lines of black and white. The feet and bill are greenish-yellow. It makes a nest in the sedges, reeds, and rushes, and lays six or seven clay-brown or gray-brown eggs, spotted with dark brown; they are from seven to ten in number. A sitting of eggs taken in Suffolk is in my possession.

Ballon's Crake (*G. Ballonii*) is a very rare British bird. It is the smallest of the European species, being $7\frac{1}{2}$ inches long, and $18\frac{1}{2}$ inches in expanse of wings. It is olive-brown on the head and yellowish on the neck. The wings are nearly black, the feathers being margined with olive; the rump and tail olive-brown; the chin, throat, and sides of the neck ash. It is said to have bred in Britain; the eggs are a nearly uniform olive-brown.

The Little Crake (*Grex pusilla*) is rather a heavier bird than "Ballon's;" but although a summer visitor, is too rare to be worthy of extended notice.

The Water-Rail (*Rallus aquaticus*) is 10 inches long, and 15 in expanse of wing. There is little difference between the sexes in colour; the bill, which is reddish-yellow at the base, and dull colour at the point, is paler in the female; it is about an inch and a half long. The iris is yellow-red, and the feet are reddish-brown; the toes are long. The upper part of the bird is olive-brown; the feathers are nearly black in the centre; the plumage of the lower part is marked with black and white on the thighs and sides. The wings are dusky, with some white marks. The wing has a horny projection. The tail is short and broad, brown and black above, with white underneath. The nest is made of the leaves of water-plants, and is placed among the thick bushes by the side of the water. The eggs vary in number; the female usually lays seven or eight; some have a white ground, with lilac and red spots, others have darker spots; a few lilac spots are almost always visible. It is found in Suffolk, Norfolk, and other eastern counties mostly, but is occasionally seen throughout England. The young of the water-rail are black, without any bright colouring.

WATER-RAIL.

The Moorhen (*Gallinula chloropus*, family *Gallinula*) is known by a variety of names. It must not be confounded with the female grouse, which in Scotland is called the moorhen. It is not a migrant, and is generally distributed. It is a good swimmer, but is heavy for its size, which is that of a pigeon, for it weighs a pound. The colour of the bird is dark olive-green, and blue-gray on the lower part, with white under the tail and the edges of the wings. There is a white spot under the eye. The feathers hang loosely over the thighs; these are black and white. The bill is thick and arched at the tip, it is greenish and red at the point; it is only about an inch long. The colour of the upper part of the bill, which forms a kind of plate on the forehead, is bright red during the breeding season, but afterwards becomes pale. The iris is red. The feet are not webbed, but the toes are bordered with membranes. The feet are green or yellowish, the toes are very long. It frequents fresh water, especially slow-running streams and deep ditches which are bordered with rushes and sedges. It is partial to the neighbourhood of man's dwellings, and comes out to feed chiefly at night or late in the evening. It feeds on small fishes and other inhabitants of the stream or pond. It lays its eggs so near the water that they are liable to be washed away; and the young birds, according to Mudie, have an enemy in the heron, which will swim after them, although this habit is not usual with the species. It is even said that in water where pike or trout are found these fish, as a compensation perhaps for the many small fry devoured by the water-hen, will devour the young birds from the water brink. They are soon able to swim; the mother takes them out early, and brings them back to the nest during the middle of the day, and when evening comes on. There is often more than one brood in the year. The eggs are seven to eleven each time, and the female sits about three weeks; the colour of the eggs is clay-brown with spots of red and cinder-gray; they are very variable; they have a fine gloss; they are rarely pointed. The nest is formed of the stems of the bulrush or sedge, and is a loose structure. The young when first hatched are of a beautiful black, ornamented about the head and face with red and blue.

The Collared Pratincole (family *Glareolidæ*, *Glareola torquata*) is a bird of most rapid flight, and has been but rarely known to visit Britain; in Eastern Europe, where it is common, it is a marsh-bird.

The Coot (*Fulica atra*, genus *Fulica*) breeds on almost all our larger ponds and lakes throughout England. It is about 18 inches in length, and 28 in expanse of wings, and weighs from a pound and a half to two pounds. It is of a sooty-black colour, and has a peculiar bald patch on the forehead; hence the name, "bald coot." Its toes have lobed membranes, which enable them to swim better than nakedfooted birds. The claws are very sharp, and it is somewhat unsafe to handle when wounded. It feeds on seeds and roots, as well as on shell-fish, spawn, and water insects. The coot is a most clumsy walker. Its eggs have the same character as those of the moorhen, but are much larger, and have the spots blacker. The nest is more cup-shaped.

The Red-necked Phalarope (family *Phalaropidæ*, *Phalaropus hyperboreus*) is a pretty little bird which breeds in the fresh-water lakes; it lays four eggs of greenish clay colour, thickly spotted with black, cinder-gray, and red; like the coot it has lobiated feet, but it is not a quarter of the size of that bird. It is 8 inches long, and the expanse of the wings is 14 inches. The plumage of the head, nape, cheek, and sides of the neck, is ash; the back, black; a white bar is across the wing; the front and sides of the neck, reddish-brown. This is the breeding plumage of the males. In winter, the black is brown, and the brown buff. It is common in winter.

The Grey Phalarope (*Phalaropus lobatus*) is an exclusively winter visitor, and a very rare one.

The divers, *Colymbidæ*, and the grebes, *Podicipæ*, from their peculiar construction, are the best swimmers and divers amongst birds. The pointed beak, the whole body tapering towards the head and tail, the powerful muscular development of the legs and feet, which in the grebes resemble a paddle of three blades, enable them to pass through the water at an amazingly rapid rate; often have we watched them as they appeared on the surface of the water to breathe, and then suddenly dive and re-appear on the water at such a distance that we were astonished at the rapid progress they had made; they use their wings as paddles.

The Red-necked Grebe (*Podiceps rubicollis*) and the Horned-Grebe (*P. cornutus*) are accidental visitors to Britain, and being similar in habits to our resident species, will not be here mentioned in detail.

The Eared Grebe, according to the "British Miscellany," page 18, remained to breed, in June 1805, on Chelsey Common.

The Crested Grebe (*P. cristatus*) breeds on freshwater pools and lakes of our country; it is most common in those of the eastern counties. It forms its nest of reeds and dry aquatic plants on the margin of the water, but not unfrequently so near that its nest is detached by the rising water until it floats. The eggs are four in number, usually a long oval pointed at each end. They are of a pale blue colour,

GREBE.

coated over with a chalky crust which usually obscures the colour. They are stained by the nest in various ways, being sometimes quite yellow. The grebe is a shy bird, difficult to approach, but is carefully sought for, for the sake of its breast feathers, which ladies fancy as an article of dress. It is 22 inches long, 30 inches in expanse of wings, and weighs from two to three pounds.

They destroy large quantities of fish and their ova, as well as shrimps, prawns, and young crabs.

The Little Grebe (*P. minor*), known also as the dab-chick, is abundant on fresh-water lakes and still-flowing rivers throughout the country. It is one of the most interesting of our little aquatic birds, and is an ornament to a piece of water. It forms a cup-

LITTLE GREBE.

shaped nest of aquatic plants. The eggs are not larger than those of a pigeon, and are stained similarly to those of the crested grebe. Grebes cover their eggs in order to conceal them. The crested grebe has the bill longer than the head. The beak is reddish, with the tip white, and, contrary to the usual habit with birds, the beak becomes of a deeper red after death. The face is white; the crown of the head, the crest, and sides of the cheeks a bright black, which shades into reddish on the sides of the head; all the lower parts of the bird are of a silvery white, the upper brown and blackish. The secondary feathers of the wings are pure white, slightly reddish where the wings join the breast. There is a naked space from the corner of the beak to the eye, the iris is crimson; the feet are blackish, the under part yellowish-white. The length of the bird from the point of the beak is 18 or 19 inches. This is the plumage after the third moult. The female does not differ much from the male, the crest is shorter and the colours duller. The crest does not fully appear till the bird is three years old, but a short crest and fringe of feathers at the sides of the head are developed after the second moult; also a black band is visible under the eyes. Before moulting the young have the head and neck of a deep brown, and instead of the crest and fringe which afterwards appear, have lines of blackish brown in zigzag form. The iris is then yellow.

The bills of the grebes are straight, hard, and compressed; the nostrils at the sides of the beak are oblong, closed at the back by a membrane and open in the front; the feet are long, placed far back, three toes before, one behind, the fore toes are much spread, but united at the base, and surrounded by a membraneous fringe, as is the back toe; the nails are large. They have a very short tail, and wings not long. The food of the grebes is fish, amphibia, winged insects, and vegetables. They walk very awkwardly, on account of the way in which their legs are placed; they stand nearly upright when on the ground. The length of time they take to attain the full adult plumage has caused them to be given a variety of names, and has led to a confusion of species. The feathers lie very close, and are very silky and lustrous; underneath the feathers is a thick down, which also lies very much compressed.

The Northern Diver (*Colymbus glacialis*) is in general a winter visitor to this island, although a few have been obtained in the summer plumage. It seems probable, therefore, that it occasionally breeds in Orkney and Shetland, although its eggs have never been satisfactorily collected there. It is about 2 feet 6 inches long, 5 feet in expanse of wings, and weighs from twelve to fourteen pounds. It is not much less powerful as regards wings and size of body than the larger birds of prey. When migrating, it flies so high as to get beyond the reach of sight. The bill is strong, and above 4 inches in length; the upper mandible bends slightly down, and the lower, which is hollow and deeper in the middle, bends up. This form of bill is characteristic of the grebes and divers, and enables them the more easily to raise their prey out of the water. In the northern diver the head and neck are black, with two white bands round the neck, marked with small black lines. The whole of the upper part of the bird is ornamented with white marks like snow-flakes, on a black ground. The rump and tail are beautifully marked with black and white; all the under parts of the mature bird are a beautiful white. The young are brown and gray instead of black and white, and the collars round the neck are indistinctly marked until the third year. The bird lays only one egg, dark olive-brown with black spots; it is about the size of those of the domestic goose. It utters a loud, discordant sound. It is perhaps the finest and most imposing of the water-fowl, whether we consider its size or the beauty of its markings. The feathers lie very thick and close, and under the skin is a layer of fat of unusual thickness in sea-fowl. Thus protected it enjoys a uniform temperature, which is fitted to maintain its large body in health, exposed as it is to the cold of the Northern sea.

The Black-throated Diver (*Colymbus Arcticus*) is a rare visitor to Britain. It breeds in Sutherland on the fresh-water lakes. "It was in the summer of 1834," says Mr. Hewitson, "during an excursion to explore the zoological productions of the county of Sutherland, that Mr. Selby was rewarded by discovering this beautiful species breeding upon most of the inland lochs. The eggs, which are two in number, were first found upon a small islet at the bottom of Loch Shin; they were upon the bare ground, and at about 10 or 12 feet from the water's edge. The bird was observed by Sir William Jardine, who formed one of the party, whilst in the act of incubation, and was seated upon its eggs in a horizontal position, and not upright like the cormorant, shag, and guillemot; and whilst still in this position it thrust itself forward when disturbed, and had thus worn with its breast a distinct track to the margin, whilst launching itself boat-like into the lake. We did not see this species in Shetland, nor is it met with in Orkney, and much to our disappointment and surprise it was only once that we got a sight of it during the whole of our journey round the Norwegian coast."

The black-throated diver is a much smaller species than the great northern diver, its length is from 2 to $2\frac{1}{2}$ feet, and the expanse of wings 3 feet to 3 feet 8 inches; its form is more slender than that of the great northern diver. There is a stripe along the fore-neck and the side of the neck, the scapulars and covert feathers are spotted white, the back and tail are black, the head gray, the feet are dull brown, and the under part white. In the young, as in the great northern diver, the full colours do not appear for three years, brown predominating until the second year, when the head becomes gray and the black marks appear; in the third year the white and black marks are seen on the various parts. The birds in all these stages seen together have been mistaken for different species. Its habits much resemble those of the great northern diver.

The Red-throated Diver (*Colymbus septentrionalis*) has the sides of the head and throat of an ashy-red, or mouse-colour, the crown is marked with black spots, long white and black stripes and a band of chestnut-red are on the neck, the breast and lower parts are white. The back is blackish-brown on the old birds, but in those of three or four years there are small white spots on the brown. The bill is black, the iris is brown-orange, the feet are outwardly dark green, underneath they are of a livid white. The length is 23 or 24 inches.

The young are at first of a brown-black on the upper parts, and whitish on the lower. At the first moult the space between the eye and beak, the sides of the neck, are white, and the top of the head is ash-gray, sprinkled with small spots of white, the coverts of the wings edged with white, the beak ashy-white, the iris brown, the feet brown, underneath ashy-white. At the end of the first year the young begin to have the same colours as the old birds, but sometimes the neck is white, with a few chestnut-coloured feathers mixed; the white spots become less distinct on the upper parts, and are sometimes yellowish. After the second moult the front of the neck is reddish, with a few white feathers; the white gradually disappears as the bird gets older.

This species inhabits the English, French, and Dutch coasts in winter, and the young are common in the inland seas of Holland, in Germany, and even Switzerland and Italy. They lay two eggs, the same shape at both ends, of an olive-brown, with a few black spots. All the divers form their nests of seaweed, or decaying marsh plants, and cover their eggs when they leave them. The red-throated diver breeds on many parts of the mainland of Scotland, also in the Orkney and Shetland Isles.

The Ducks and Geese (*Anatidæ*) are *par excellence* birds of fresh water. Of the Gray-leg Goose (*Anser cineraceus*) Mr. Wolley, as quoted by Mr. Hewitson, says: "I believe it to be the only species of goose which has been in the habit of spending the breeding season in Great Britain, formerly in great numbers in the southern part of the island. During several seasons I have made particular inquiries of the most competent persons at these breeding-places in Sutherlandshire, where the bean goose has been said to be found; and the descriptions of the only kind of goose known by them to breed in the country have always been referable to the gray-leg. Everywhere I examined the bird with a glass, and in several instances shot specimens, and all I saw or obtained were gray-leg. The gray-leg chooses various kinds of places for its nest. I have seen, in Sutherland, nests on the open moor, but not very far from a loch, and again within 2 or 3 feet of the water's edge; but generally they are on islets in the sea or in fresh water. It would be difficult to find anything more beautiful than the little islets in some of the Highland lochs to the lover of nature in general, but to an ornithologist they are surpassingly so. None have made greater impression upon me than two on a retired piece of water in Sutherlandshire; they were very small, rising up somewhat steeply, and were covered with long heather and other plants bedded in the most luxuriant moss. In each were two or three little trees, and in each was a huge nest of the sea eagle, fixed so near the ground that a child could see into it; one nest some years old, the other repaired that season. Hooded crows built in the branches over the newer nest; and in spite of the frequent visits of the eagles, a wild duck had its nest not many yards off, and geese bred there regularly. The other islet had been burnt several years before to dislodge a fox, and now its bright, young heather again formed an excellent cover. A pair of black-throated divers crying on the surface of the loch, two wild geese flying round, and an old eagle, with its broad, white tail, slowly wafting its way between me and the neighbouring mountain, whilst the great nest was conspicuous from every side, made it altogether as delicious a scene as I could hope often to enjoy. I had not walked many paces when a gray goose fluttered from between my feet into the water, not looking at all a large bird, and not getting up with any great commotion. There were at present only two eggs in a nest made of old, withered grass, like others which I afterwards saw."

The Shieldrake (*Tadorna Vulpanser*, sub-genus *Tadorna*), usually a sea-shore duck, is sometimes, however, found on the estuaries of rivers, lakes, and pools; it is 2 feet long, and $3\frac{1}{2}$ feet in expanse of wings, and weighs two pounds and a half to three pounds. The feet and bill are reddish, the head and neck are green, the lower part of the neck and upper tail-coverts are white. There is a band of fine red-brown on the breast, which forms a narrow collar on the lower part of the neck. The female is like the male in plumage; it nestles in holes, where it lays ten to fifteen eggs, of a cream-colour, larger than those of the common duck.

The Shoveller (*Anas clypeata*, sub-genus *Rhynchaspis*) is easily distinguished from all other British ducks by its shovel-shaped bill, and bright light blue on the wings. It is much smaller than the Shieldrake, and is excellent as an article of food, which is generally the case with ducks which, like it, live on fresh water. Mr. Hewitson says, " Mr. John Hancock has the nest and eggs of the shoveller, which were found upon Prestwick Carr, a piece of waste ground of considerable extent, near Newcastle-upon-Tyne, covered with heath and furze, boggy and intersected

SHIELDRAKE.

with drains, and having a piece of water near its centre. From thence, towards the end of May, a nest was brought to him containing nine eggs; it was composed of grass, mixed with the down of the bird, and was placed in the centre of a furze-bush, by which it was sheltered. Two or three weeks after this a second nest was found at a short distance from the spot from which the other had been taken. It was constructed of the same materials, was situated similarly, and contained ten eggs; these were quite fresh, and led us to suppose that they belonged to the same bird which had been previously deprived of its eggs."

The Gadwell (*Anas strepera*) is a winter visitor to our inland lakes. It is about 19 inches long, and 33 inches in expanse of wings; it is marked with exceedingly minute bands on the head and neck of a crescent form on a gray ground, the scapulars and flanks with zigzag lines of black and white.

The Common Wild Duck (*Anas boschas*), believed to be the source of most of our domestic breeds, which, like those of Gallinaceous birds and pigeons, are subject to an almost endless variety of colour and form of plumage. The nest is usually found on the banks of a stream or lake, sometimes on the ground, at other times on the top of a bush or tree. Mr. Tuke met with it 20 feet above the ground; Mr. Tunstall, 25 feet from the ground; Mr. Selby, in the nest of a crow, at least 30 feet from the ground. The nest is round, and is lined with white down, the exterior being made of dry grass. Its diameter inside is about 6 inches, but the down is 3 inches thick. The eggs are about eleven in number.

The Pintail Duck (*Anas acuta*) is another winter

visitor to Britain; it is distinguished by long tail-feathers.

The Teal (*Anas crecca*), although more abundant in winter, is a breeder on the margins of our lakes.

"Prestwick Carr," says Hewitson, " a fine piece of wild moorland, intersected in all directions by drains and spongy swamps, a few miles from the town of

TEAL.

Newcastle—well known by the naturalists of the neighbourhood for its riches in each of the branches, and probably frequented by a greater number of species of birds than any place of a similar size in this country—is one of the breeding-places of the Teal. In Mr. Hancock's collection are two nests of this species, taken by himself on the 28th of April, and each containing eleven eggs, the full number; they were placed among the long heather, of which, together with some dry grass, they are outwardly constructed, and lined, 2 inches thick, with the softest down, kept together by having bits of heath and the stalks of grass interwoven with it. One of them is a very beautiful object, each separate piece of the down with which it is lined being outwardly of a dark brown with a pure white centre. Mr. W. M. Tuke has found the eggs of the Teal on Strensall Common, an extensive waste near York, and very much like the one I have just described; they were placed, without any nest, under the shelter of a piece of furze."

The Garganey, or Summer Teal (*Anas querquedula*), has been said to breed on the pools of the South of England, but it is doubtful, though it has certainly been often killed in the breeding season. It is a fresh-water species.

The Wigeon (*Anas Penelope*) was discovered by Mr. Selby, breeding in Scotland, who wrote to Mr. Hewitson as follows: "The nest from which the eggs were taken was upon the island in Loch Laighal, upon which is a large colony of the lesser Black-backed Gull. It is covered with ferns and other long herbage; and the nest, well-concealed in a thick bed of rushes, was composed of their decayed stems and other grasses, with a large quantity of the bird's down interwoven, the eggs being far advanced, and the young nearly ready for extrusion. The female we shot when she arose from the nest. Upon most of the rocks were several pairs."

The Common, the Velvet, and the Surf-scoters are marine ducks, but are occasionally found on the estuaries of rivers, when driven there by stormy weather. The same may be said of the Scaup Ducks; the Eider, likewise, is a sea species.

The Pochard (*Anas ferina*) is one of our best ducks for the table, and is extremely beautiful. It is 20 inches long, 30 inches in expanse of wings,

POCHARD.

and twenty-nine ounces in weight. Mr. Hewitson thus describes its breeding in this country on the information of Mr. James H. Tuke :—" Whilst at Scarborough, about the middle of June, Mr. Bean informed me that several pairs of red-headed ducks as the gamekeeper called them, had been seen upon a piece of water a few miles from Scarborough, and that he was going the next day to see if he could find their nests. I had the pleasure of accompanying him, and, sure enough, several pairs of pochards flew up from their reedy habitations as we passed our boat up amongst the tufts of grass and long reeds, which at one end of the lake form a bog of many acres in extent, almost inaccessible, for between these tufts of treacherous grass the water is some feet deep. Is was with the greatest difficulty we managed to jump from one of these tufts to another. While beating about amongst this herbage, a female pochard flew up almost close to us, and in a short time the gamekeeper, who was with us, found a nest lined with feathers, and rather under the shade of a bush of *Myrica gale*, which grows plentifully in this bog. I had the pleasure of seeing the nest, but unfortunately there were no eggs. After trying in vain to find another nest, we marked the spot, and left. Mr. Bean returned in a few days, and found eggs in this and another nest very near it, from which the one I sent you was taken."

The Red-crested Pochard, a still more beautiful species, is likewise a bird of the fresh-water lakes, but is an accidental visitor, as are the Tufted and the White-eyed Ducks.

The Long-tailed Duck (*Anas glacialis*) is a winter visitor in considerable numbers; it frequents the sea-shore and estuaries of rivers.

The Scaup Duck (*Anas marila*) was discovered by Mr. Selby, with a young one, on a small loch in Sutherlandshire; proving that it breeds there.

The Red-breasted Merganser (*Mergus serrator*, genus *Mergus*) was found by Mr. Selby and Sir William Jardine breeding on the shores of Loch Awe, as also by others in the Hebrides, Shetland, and Ireland. The nest is placed among grass in a hole scooped in the earth, forming a perfect circle; the eggs are six in number, and grayish cream-colour.

LONG-TAILED DUCK.

The Red-breasted Merganser (*Mergus serrator*) is one of the most beautiful of our ducks. It is 20 inches long, 30 inches in expanse of wings, and with a beautiful crest on the head. The bill is long and toothed, which enables it to seize fish with great facility; the head and upper part of the neck a beautiful green, with a purple sheen. The female is much plainer in plumage.

The Goosander (*M. merganser*), the Hooded Merganser (*M. cucullatus*), and the Smew (*M. albellus*), are sometimes found on our larger sheets of inland water in winter, although, properly speaking, oceanic birds.

The Gulls (*Laridæ*) are seldom, comparatively speaking, seen on our fresh-water lakes or rivers, with the exception of the Black-headed Gull (*Larus*

THE BLACK-HEADED GULL.

rudibundus), which breeds on the ponds and lakes of our eastern counties. Mr. Hewitson says, "The most numerous colony which I have seen occupies a piece of water upon the estate of Mr. Askew, at Palingsburn, in Northumberland, being little disturbed." At most of their breeding-places they are less fortunate, for the eggs are much in request, having a good flavour and no fishy taste, for the birds at this time of the year live on land slugs and worms chiefly. When the eggs are removed, like many other species, they lay again, and even a third time, but the eggs are smaller, some not above a third the proper size. Some eggs of this species are as abnormal in shape and size as eggs of the common fowl frequently are.

"If we adopt," says Mr. Hewitson, "the opinion of some naturalists, that the ovarium of a bird contains, from its first creation, all the eggs which it is destined to lay through life, then how soon must those persecuted gulls be rendered barren and unproductive, perhaps even before they have once had the pleasure of bringing up a family of young ones." This species breeds in Norfolk, and especially at Scoulton Mere, where there is a large colony of them. They perch upon the willow-trees in an ungainly way. Mr. Newton says he saw a nest in one of these trees, placed about 4 or 5 feet from the ground. The eggs are laid in the end of April or May. Quantities of these eggs are brought to Norwich market. A man and three boys will collect as many as a thousand in one day: this shows how extremely prolific the birds are, and that in this locality they assemble in great numbers. The eggs vary very much in colour; the ground is sometimes a light blue or yellow, and sometimes green, red, or brown.

CHAPTER V.

AMPHIBIA.—FROGS, TOADS, AND NEWTS.

THE reptiles which inhabit our rivers, ponds, and ditches, and which are all amphibious, are not of numerous species, though very abundant and widespread. Of these the common Frog (*Rana temporaria*) is the most abundant, and from its many transformations is highly interesting. Most persons must have observed from time to time gelatinous masses lying about the edges or among the reeds, rushes, and water-plants of our common pools: these are the spawn of the common frog. In each of these eggs which form the mass there is a small black spot, which is the commencement of the life of a young frog or tadpole. As there is no cessation in the progress of Nature, whether it be in life or in decay, development commences from the time of deposition. Soon the head of the tadpole may be discovered, and a flat tail becomes visible. Afterwards gills or fibrous tufts appear on the sides of the neck; these temporary appendages float loosely in the water, and take from it the oxygen, which is the breath of life to the young creature. The development and changes in the structure now proceed rapidly. Five days after being spawned, the tadpole, with its gills and tail, approaches its full length.

Small arms or holders have been developed near the mouth, or the part which is to form the mouth. At first the mouth is small, and the nostrils and eyes are not visible, but soon they begin to show themselves. The bronchial tufts or gills grow, and are divided into lobes, but the gelatinous envelope is not entirely got rid of; this the young tadpole shortly disposes of by its struggles, and the tadpole stage arrives at its maximum. The circulation of the blood in the gills or branchiæ can be well seen if the animal be placed under a microscope. These parts are so transparent that the globules of the blood, when highly magnified, can be perceived. A change takes place in the branchiæ; gradually they are transformed into more perfect gills, four on each side, covered by a membrane. The mouth and eyes increase in size, and the filaments of the sides of the mouth disappear; the tail is now so expanded that the creature can swim about in search of its food.

But although we are describing a single tadpole, yet there are many in the same stage which have been produced from the numerous eggs or spawn, and now form a shoal of greater or less extent, numbering perhaps hundreds. Further changes take place, hind legs appear, at first under the skin near the tail; fore legs also. Simultaneously the gills gradually disappear, for the lungs, which were in embryo within the tadpole, expand, and it begins to require atmospheric air; as the animal becomes more terrestrial and less of the nature of an aquatic creature, the tail diminishes till it finally disappears, the legs acquire their full proportions and strength, and it can either leap upon the ground or swim in the water. The spinal column of the frog is extremely short, consisting of "eight vertebræ, exclusive of those which have united to form the *os coccygis*. It was evidently the design of the Creator to consolidate the framework of the trunk, in which flexibility was not required for progressive motion, the performance of that function being transferred to the hind extremities, which are exceedingly large in proportion to the rest of the body. In every part of the skeleton there is a ten-

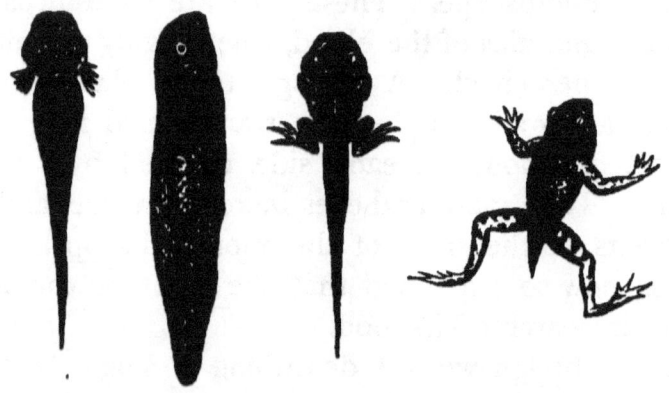

THE DEVELOPMENT OF THE FROG.

dency to develop itself in a transverse direction, while the trunk is shortened as much as possible. The mode in which the vertebræ are articulated together approaches that of the higher animals."

Dutrochet found, by careful study of the spawn and formation of the cartilage first, and then from the formation of the bones, that before the cartilaginous coccygeal vertebræ had ossified, the tail disappeared; "the vertebræ nearest the body became consolidated, and being joined to the sacrum at an angle, gives rise to the strange appearance of that part of the back of a frog which *in the skeleton* looks as if it had been broken. At the same time the spinal cavity (as in the tadpole) is obliterated, and the spinal marrow, which the tadpole stage had passed through, is withdrawn."

The anatomy of the frog is simple; the ribs are only small, detached bones, or rather cartilages, affixed to the extremities of the transverse processes of some of the vertebræ. The bones of the arms and legs resemble those of the higher mammalia. There are five toes on the hind foot, and sometimes the rudiments of a sixth. The fore foot has only four toes, without claws. The hind feet are adapted for striking the water backward in swimming; while this formation causes an awkwardness in walking on land, it greatly facilitates the long leaps which are characteristic of it.

The tongue of the frog is greatly used by it for catching the insects which are its prey. On the tongue is a viscid secretion which fixes small objects when once they touch it, thus greatly helping the animal in feeding. The frog hybernates in the mud of the pond or stream it frequents during the cold months, and in spring emerges to commence the office of propagation. The frog is extremely useful in the garden in summer, devouring quantities of slugs and the larger insects, which are injurious to vegetation. The extraordinary and sudden appearance of frogs and toads in places where they had not been known to be present is among those wonders which have not hitherto been satisfactorily accounted for. Frogs can be tamed, and have been known to frequent the kitchen, and often re-appear at meal-time, familiarly seeking for their share of provision. The Edible Frog (*R. esculenta*) is larger than the common frog, and is green, with olive-green spots, and light-coloured marks on it. It is much eaten in some parts of France, but in other localities as much despised as an article of food as it is in England; the flesh is said to resemble that of a chicken. This species is found in a few localities in England, but is supposed not to be indigenous. There is a species of frog called the Scottish Frog (*R. Scotica*), mentioned by Bell, which is local. The common Toad (*Bufo vulgaris*) is an animal of evil repute. This feeling is so ancient and widespread that we cannot but think there is some foundation for it. Yet naturalists, and those who have most considered the subject, generally call it a harmless animal. The toad is inactive and sluggish, yet, according to Mr. Bell, it feeds only on insects which are in motion. Like the frog, its tongue is furnished with a viscous secretion; remaining quite still, the toad watches for its prey, and when within reach, with the quickness of thought, darts out its tongue, catches the fly, or whatever it may be, and secures it irrevocably. The venom of the toad, if such it be, lies in the thick yellowish mucus which lodges in the pores and warts of its skin: when disturbed, it exudes a considerable amount of this liquid. The toad is said to breathe partially through this porous skin, which is always moist, and this is required for the breathing function, just as in fishes, when the gills are dry, the office of breathing becomes intercepted. The colour of the toad is brownish-gray above, and light-yellowish underneath, sometimes spotted with black; the warts on the back are dark in colour. Toads are said to be very long-lived, and to be able to do without food for many years shut up in the middle of a stone, from which even the air is excluded. That toads and other reptiles are extremely tenacious of life, and can exist under circumstances which would be fatal to animals of a higher class, may be readily admitted; but the difficulty of proving to absolute certainty the length of time and all the conditions under which a toad has existed for years is so great, that it is extremely difficult to decide positively on the truth of many anecdotes which are related.

The Natter Jack Toad (*Bufo calamita*) is another British species, which is local, and getting scarce. It is able to endure greater drought than the common species, and is more active and less sluggish in its motions. It is often found on heaths and dry places, proving that it can live a considerable time without water. It much resembles the common toad, except that it has a line of yellowish colour down the back; its colour otherwise is light brown, with patches of a darker hue. The late Dr. Buckland made many experiments on toads. In order to prove how long they could live without air he enclosed several in sandstone. He found that full-grown toads lived in this situation for nearly two years, but that the younger ones died much sooner. They were placed with a glass aperture in the stone, so that they might be from time to time examined without the admission of air. He also enclosed four toads in the trunk of an apple-tree in spaces which were cut out for the purpose, being afterwards carefully plugged up. At the end of a year the holes were opened, and the toads were found dead and putrified. Dr. Buckland, from these and other experiments, drew the conclusion that toads cannot survive a year without air, and probably cannot live two years without food.

Starvation is a cruel death, even to an animal so humbly constituted as a toad, but we may not suppose that its sufferings are in proportion to the time required before its existence comes to an end.

The custom of rearing and maintaining Newts and other amphibia in aquaria has given opportunities for observing their habits to many persons, who would not, perhaps, have had the patience and persevering watchfulness required to seek them out and study them in their native haunts. These creatures are often called Tritons. The tail is flat at the edges, and there are no parotid glands at the sides of the head. The body is covered with warty excrescences, and the male has during the breeding season a membranous crest along the back, and another along the tail; but after this season the crests get smaller and almost disappear. The newt lives almost entirely in the water, and swims chiefly by means of its long tail, which it moves to and fro like a paddle, for the fore feet are short and small, and have little power comparatively in impelling it rapidly either in the water or on land. There are four British species. The common Warty Newt (*Triton cristatus*) is about 6 inches in length; it is found in ditches and ponds almost everywhere. It feeds on insects, the spawn of frogs, and even smaller specimens of its own kind. It is a fierce little creature, and holds and devours its prey with eager greediness. As spring approaches the crest in the male grows, and becomes deeply notched or jagged, looking as if cut out in the form of a frill. The female deposits her egg singly on a leaf, and with her hind legs draws the edges of the leaf together, which having become coated with the sticky slime which exudes from her body, are glued together, enclosing the egg, and protecting it securely. Mr. Bell says that he has caught with a minnow net many of these reptiles in the ditches in the neighbourhood of London. In May or June the female deposits her eggs, and the young or tadpoles, with fringed gills, swim about. Towards autumn the gills disappear, and they acquire their perfect condition. During winter they hybernate in the soft mud under the water. They are capable of sustaining life during a great amount of cold, remaining torpid in frozen water, and are even said to exist when enclosed in blocks of ice itself, and to return to activity when the melting ice sets them at liberty. A remarkable peculiarity prevails in tritons; it is, that they have the power of renewing their limbs, and even the tail, if it should be broken or cut off. This is said to be repeatedly performed by the same individual, and is a physiological peculiarity which only occurs in the lower orders of animal life, and is, of course, quite unknown and impossible in any of the higher classes; it is analogous to the power which crustaceans, such as the lobster, have of reproducing their claws and shells, the power of repairing damages which accidentally or otherwise may happen to them. The other three species of newts mentioned by Mr. Bell in his classification of British amphibia are the Straight-lipped Warty Newt (*T. Bibronii*), the common Smooth Newt (*Lissotriton punctatus*), and the Palmated Smooth Newt (*L. palmipes*). These newts, though, in common with all reptiles, repulsive to most persons, yet are interesting and pretty when observed in an aquarium, and the remarkable changes they undergo make them among the most curious and interesting in the wide field of lower animal life.

THAMES FISHING BOAT.

CHAPTER VI.

FRESH-WATER FISHES.

SCIENTIFIC writers separate Fishes into two comprehensive divisions, Osseous and Cartilaginous. The skeleton of osseous fishes, as the name indicates, is bony and hard; in cartilaginous fishes there is no hard bone. The backbone, so called, is gristle, and if boiled becomes gelatinous. Osseous fishes consist of soft-finned and hard-finned. Order 1 is the *Acanthopterygii*, or Spinous-finned fishes.

THE PERCH.

The Perch family, or *Percidæ*, comes first in this class. Only ten or eleven species are found in Britain, although Cuvier in his classification of the fishes of the world enumerates 500 species of the Perch family. The body is oblong, and covered with hard scales, and the bones of the cheeks are serrated or spinous; the jaws and throat are studded with teeth. *Perca fluviatilis* is as well known as most fresh-water fishes, and is considered good for the table. This fish is tenacious of life, and will live for a considerable time out of water. A perch weighing 2 lb. is considered a large fish, though they are taken sometimes of greater size. The perch is extremely prolific; according to Mr. Yarrell 280,000 ova to the half-pound weight have been found in a perch. Its bright colours make it one of our most beautiful river fish. Couch says the *Serranus cabrilla*, which used to be thought of as exclusively a Mediterranean fish, is found in the rivers of the West of England. It has a habit when dying of keeping its fins erect, and its mouth open, from which it has obtained the name of the gasper. The Pope (*Perca fluviatilis minor*) is about 4 or 5

THE POPE.

inches long. It inhabits canals in the neighbourhood of London and elsewhere.

As exemplifying the power which this fish has of sustaining life for many hours when out of the water, it may be stated that perch on the Continent of Europe is sometimes taken from the pond to market in the morning, and, if not sold, brought back and replaced in the water at night. They go in shoals, and are very abundant in this country, for most rivers and clear pieces of water contain them. The Perch prefers a rapid stream of clear water, but still it is often to be found in ponds and canals, which are less wholesome and enjoyable to it. "Mr. Jesse said some years ago that great numbers of Perch were bred in the Hampton Court and Bushey Park ponds, which are well supplied with clear water, but they seldom grow large."

In the Regent's Canal they are said to be very abundant. The full-grown Perch of this canal is said seldom to weigh more than from half to three-quarters of a pound, and that those taken have a surprising uniformity in size and weight. Izaak Walton advised the angler to commence fishing for Perch when the mulberry is putting forth its buds. He describes it as a bold-biting fish, not deterred from coming to the enemy by the loss of many of its companions.

There are four species of the genus *Cottus*, but only one is a fresh-water fish.

The River Bullhead (*C. gobio*), vulgarly called the Miller's Thumb, is about 4 inches long. It hides itself among loose stones, but it is a quick and active little fish; it is very common in almost every stream. It is also called "Tom Cull" in some localities. The head is very broad and the mouth large. Of the genus *Gasteroteus* is the three-spined stickle-back, Banstickle, or Sharplin (*G. spinulosus*); it is 2 or 3 inches long. This fish has a curious habit of building a nest. It collects little pieces of stick and fibres, which it fastens together, leaving the inside hollow for the reception of its eggs. The task of making the

BULLHEAD.

nest and depositing the eggs is said to occupy one day only, and the male fish watches assiduously over the eggs and young fry. About a month intervenes until

GREY MULLET.

the young emerge from the nest. The Grey Mullet (*Mugil capito*) inhabits estuaries on the coast, and in summer is found in muddy bottoms of rivers, so that it is partly a fresh-water fish. It is well known as a delicacy for the table on the south coast. They are often taken with the rod, either by fly or worm; they are from 15 inches to 2 feet in length, and are steel-gray in colour, with reddish-brown bands; the belly is silvery white.

The second division of osseous fishes includes the *Cyprinidæ*, or Carps. These have usually the fin-rays flexible. *C. carpio*, the common Carp, has the tail-fin forked, and the mouth has two barbules on each side. It is a broad and thick-shaped fish; the colour is

CARP.

olive-brown, with a dull golden tinge. They have large teeth in the gullet; they feed on insects and worms, but hibernate during most of the winter. They are extremely prolific; sometimes the roe contains more than a million of eggs. They can bear a great amount of cold, and live to a great age. Carp was more esteemed in former times in England than it is at the present day. It is now much liked for the table in France. Carp requires skill in cooking; coming from the hands of a skilful *chef de cuisine*, swimming in rich wine, flavoured with plenty of good herbs and butter, it is a dish much relished by the gourmand. In many lakes and ponds the Carp is plentiful, as it is in some rivers in England and France. This fish was much cultivated formerly in fish-preserves, and there are many remains of old carp-ponds, especially in the neighbourhood of convents. Being susceptible of conversion into a dainty dish, it was an excellent substitute for meat during Lent and other fasts. The Golden Carp, or goldfish, is now naturalized in Britain, and though formerly only known in glass vases within-doors, is now not uncommon in ponds in various parts of the country. It is a native of China, and is said by Mr. Buckland to breed most abundantly in this country in the mill-dams and ponds of the manufacturing districts, where, from access of the warm water of the steam-engines the temperature does not become very low. The common Carp sometimes weighs 20 lb., but the usual size and weight is not half this amount. The Prussian Carp (*Cyprinus gibelio*) is well adapted for ponds. It has a metallic lustre; the eye is yellow, and the fins orange-red. It weighs about a pound. Mr. Buckland says he obtained it from the Serpentine.

The Barbel (*Barbus vulgaris, Barbus fluviatilis*, or *Cyprinus barbus*) is common in rapid streams, and plentiful in the Thames near London. The name is given to it on account of its barbs or wattles at the mouth. The liver of this fish is called poisonous, and it is not otherwise a good fish for the table. It feeds on slugs and worms. It furnishes an amusing sport to those who like fishing from a punt, but as it is not desirable as food, unless they become too numerous there is not much object in fishing for them; sometimes they weigh as much as 14 lb., but in general are much less. They are gregarious. The colour of the upper

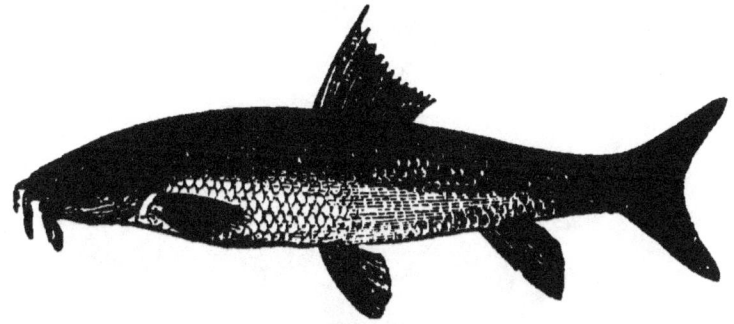

BARBEL.

part is greenish-brown, and the under light or whitish. This fish is not found in Scotland. The Gudgeon (*Gobio fluviatilis*), a small, pretty fish, which differs in

GUDGEON.

colour according to locality, is about 4 or 6 inches long, with a single barbule at the corners of the mouth. The scales are large, and there is a dark line along the body. It is found in the Thames and other rivers. When the fishermen want to catch them they scrape the bed of the river, which draws a crowd of them together. They are called very good for the table when freshly caught.

The Tench (*Cyprinus tinca,* or *Tinca vulgaris*) has

very short barbules, and the scales are small. This fish usually weighs 5 or 6 lb. It buries itself in the mud in winter. There is a curious old tradition that

TENCH.

this fish by its touch cures the maladies of other fish; even Izaak Walton believed in it, and the Pike, usually such a greedy and grasping fish, is said to avoid eating the Tench as if by instinct, on account of its virtues.

BREAM.

The Tench is considered very useful as an article of food by some writers, but it seems more in theory than practice in the present day. Like many pond fish, its flesh is soft and flavoured with the muddy bed on which it for the most part roams.

The Bream, or Carp-bream (*Cyprinus brama, Abramis vulgaris, Abramis brama*), is an inhabitant of large lakes, broads, and deep rivers. These fish feed on worms, and soft animal substances. They go in shoals, and, according to Izaak Walton, have a sentinel watching to give them warning of danger. A large bream weighs 6 or 7 lb.

THE ROACH.

The Roach (*Cyprinus rutilus, Leuciscus rutilus*) is smaller than most of its tribe, for it seldom exceeds a foot in length, and is about 8 or 10 inches long, and the weight seldom exceeds 2 lb. It is rather a stout fish. The jaws are without teeth, and the snout is rounded. It is subject to a bad disease, in which the scales become black. Its habitat is similar to that of the Bream. The Azurine, or Blue Roach (*Cyprinus cæruleus*) is a rare species in Britain. The Dace (*C. leuciscus*) is like the Roach, but smaller, and more slender in form. It is common in the Thames, in many other rivers, and in ponds. The Chub, also a common fish (*L. cephalus*, or *C. cephalus*), though despised as an article of food in England, in France, as in so many similar cases, is made into a very good dish; 4 lb. is considered a great weight for a chub, though in the Trent they are said to weigh heavier. In Cumberland it is called the Skelly. It is said by Mr. Couch to be found only in clean, clear waters.

The Rudd, or Roud (*Leuciscus* or *Cyprinus erythro-phthalmus*) is by some authorities called a hybrid between the Roach and Bream. Mr. Yarrell does not

THE CHUB.

believe in hybrid fishes. It is sometimes taken in great numbers. Its food is worms, mollusks, insects,

THE BLEAK.

and vegetables. The scales, like those of the Roach, are rough. The general colour is coppery red, and the fins are red. It is found in England and Ireland, and frequents the same kind of rivers and ponds as other members of this family. The Bleak (*Cyprinus alburnus*, or *Leuciscus alburnus*) is a small fish, dark-greenish, with a white belly. Formerly the scales of the Bleak were used for making pearls. They are said to be useful in devouring much decaying matter in the river Thames, and thus helping to purify the stream. The Bleak is from 4 to 8 inches long.

The well-known little Minnow, *Cyprinus phoxinus*, is a pretty little creature; in spawning-time the colour of the male is green and red. They are said to be good to eat when pickled, and might form a kind of fresh-water sardine. The Minnows are everywhere scattered in our pools, ponds, and rivers, and are much devoured by larger fish; yet they do not seem to decrease.

THE LOACH.

The Loach (*Cobitis barbatula*) has six barbules about its mouth. It is common; but hiding itself among the stones and gravel, is often overlooked. It is slimy and difficult to hold.

There is a smaller species, the Spined Loach or

Groundling (*Cobitis tænia* or *Botia tænia*), which is rare.

The Pike family is the second in the order of osseous fishes. The common Pike (*Esox lucius*) is called by Sir Francis Bacon the longest-lived of fresh-water fish. Mr. Buckland, in his History of Fishes, mentions a picture which he has of a pike which the Emperor Frederick II. put into the pond on 5th of October, 1233, and which was caught more than two hundred years after with a brass ring on its gills, on which was engraved : " I am the fish which Frederick II., the Emperor of the World, put into the pond, the 5th of October, 1233." The length of the fish was 4 feet 9 inches, and the ring round its neck 10½ inches. However, Mr. Buckland does not believe in this extraordinary longevity of the

THE PIKE.

Pike. We do not associate pleasant ideas with the Pike, which is a ferocious, greedy fish, the enemy of the weak and sickly or smaller inhabitants of the streams or lakes where it lies. In the days of Henry VIII., Pike was much esteemed as an article of diet; its value in the month of February was double that of a house lamb. Pike are not much in demand for the table at present. They are as prolific as almost any fishes. Mr. Buckland mentions one which he examined, which contained 595,200 eggs. Pikes fight desperately with each other.

The Whitebait (*Clupea alba*) is a problem to the ichthyologist. Many persons say they are young herring, but others say they are the fry of a variety of species; for, on examining a quantity of these small creatures, they are found very various in appearance, and evidently belong to very different species, though all called, by the fisherman and fishmonger, Whitebait. When larger fish are caught among the Whitebait, they are rejected and usually thrown again into the water.

THE WHITEBAIT.

Sticklebacks are found among the variety of young fish; those called by the fishermen "the rooshians," are young weever fish, and are like a colourless jelly. Polliwigs are the young of the spotted goby. Mr. Buckland also discovered among the Whitebait the young of other species of fish. He therefore came to the decision that there is no *real* Whitebait, but that the fish which are so named are just the fry of a variety of fish which are delicate in taste, and have not acquired the peculiarity which belongs to each species when further developed. The following extract bears on this interesting subject:

"ROCHESTER.—Messrs. Buckland and Walpole, the Sea Fishery Commissioners, held an inquiry here to-day, when several experienced fishermen were examined. Their unanimous opinion was that the catching of whitebait should be altogether prohibited, the fish to which that name is given being, in their opinion, not a separate species, but merely the fry of sprats and herrings, the herring fishery in particular suffering greatly from this wholesale destruction of young fish."—*Daily Telegraph*, Friday, November 15th, 1878.

The Twaite Shad (*Clupea alosa, Alosa finta*) is from 10 inches to nearly a foot and a half long. It has teeth in both jaws, and has a row of dark spots on each side. In former times the Shad, like many of our fresh-water fish, was more esteemed than it is now. Formerly it was abundant in the Thames.

The Allice or Allis,—the Alewisse Shad (*C. alosa, A. vulgaris*), is much larger than the above species. There is an abundance of them in the Severn and Wye, but elsewhere they are scarce.

The Flounder (*Platessa flesus* or *Pleuronectes flesus*), called in Scotland Fluke or Mayock-fluke, is of the family *Pleuronectidæ*, or flat fishes. The habit of this fish is to frequent chiefly the muddy bottoms of rivers and estuaries. It is a small flat fish, broad in the middle, and very thin. It is a pale dusky brown, sometimes spotted on the upper side, and white underneath. It swims on its side, and propels itself by waving its thin body to and fro. Both eyes in this fish are on the upper side. As the Flounder has little means of defending itself, it is compensated in some measure by keeping much to the bottom of the water, where its thin, flat body and duller colour

THE FLOUNDER.

are little perceptible. It seldom weighs more than 1 lb., though instances of much greater weight are recorded.

The family of the Murenidæ or Eels (order Apodal, or without ventral fins), contains three or four different species in this country.

The common Eel (*Anguilla acutirostris*) is so universally known as to require little description. Although it is quite a fresh-water fish, yet it, when inhabiting a river, emigrates towards the mouth in autumn, and lives in brackish water. They are caught on the Thames in their descent towards the mouth of the river by means of an "eel-back," which is the technical name for a structure of wicker baskets, supported on a wooden frame. Each basket is open at the end which is opposed to the descent of the stream. The fish, when swimming down, enter these baskets, and are unable to get out again. Londoners are very partial to this fish, which is rank, oily, and doubtless very nourishing. But it appears that large quantities of eels are imported from Holland, and have been so for at least three hundred years. They are, as is well known, extremely

EEL-BACK.

tenacious of life, and are known to leave the water and travel to a considerable distance on land. They will even climb over high objects which obstruct their progress. They have a curious pouch or bag

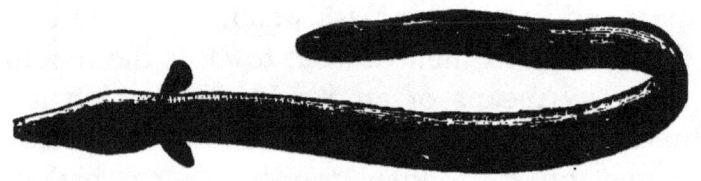

THE EEL.

containing water, which they use to moisten their gills, and in this way prolong their lives when on land. This is so remarkable a peculiarity that, were the Eel a rare fish, it would be enumerated as among the greatest curiosities of natural history.

Throughout the country, eels inhabit ditches, ponds, lakes, and pieces of water generally. Some eels are of very large size, and many instances of this are quoted in *Land and Water*, as well as in other periodicals and works on ichthyology. Six or eight pounds in weight is common. Mr. Buckland mentions an enormous eel which weighed 36 lb. It was caught in the Ouse, near Denver sluice. The back was black, the sides golden, and the belly silvery. The elvers or young eels are almost transparent, the heads are small, and the backbone can then be seen: this is the appearance of them when about a week old. They are caught when ascending the river Parrett, in Somersetshire, by the hundred thousand, perhaps million, and made into *cakes* to be cut in slices and fried, when they are said to be very good—so says Mr. Yarrell. Quantities of elvers are said also to be caught in the Severn, and similarly disposed of. Eels produce their young alive. The eel-fisheries in Ireland are very productive, and in the case of salmon culture, facilities for passing up and down the rivers have been contrived for the young eels. One method is that of placing ropes of straw in their way, up which they climb, over the objects which otherwise would be to them insurmountable.

The Broad-nosed Eel (*A. latirostris*) is also a common kind. It is called by the fishermen of the Severn the "Frog-mouthed Eel." It has a wider mouth than the common Eel, the nose is broader and the skin thicker, and it does not grow to such a large size as is frequent with the former-named eel. It is shorter and thicker in body proportionably.

The Snig Eel (*A. mediorostris*) is a small species of a yellow colour. The Hampshire Snig has the peculiarity of moving about and feeding in the daytime, which its congeners are said not to do. These little eels are superior for the table, but they are not often more than half a pound in weight. The form of the nose is less broad in proportion than the Broad-nosed Eel, and not so pointed as the common Eel.

Eels are sometimes quite golden in colour, and Mr. Buckland says he has even seen specimens nearly white, which would be, of course, albinos. This writer also states that quantities of eels inhabiting our rivers and canals are not utilized, and, especially in Scotland, where they are wasted to an enormous extent. The eel-fisheries being so productive in Ireland, it seems a mistaken policy not to make them so also in Scotland. But the Scotch have a great prejudice against eels as food, owing to their snake-like appearance, and the traditional feeling amongst them that they were forbidden by the Mosaic law.

The Snake Pipe-fish (*Sygnathus ophidian*) is a long and very small fish, not above half an inch in diameter at the thickest part, while the tail is very slender indeed. The male fish has small cells on the abdomen, in which the female deposits the eggs, and which are thus carried about by him for some time. Mr. Yarrell says they are about the size of a mustard-seed. They are often among the fry caught in rivers which goes popularly under the name of whitebait.

The cartilaginous fishes, among which first occur

the *Sturionidæ* or Sturgeons, are similar to the Sharks, having long angulated bodies with longitudinal rows of plates and spines. The snout is long and conical, and the tube-like mouth, not furnished with teeth, is under the head at a distance from the nose. Between the mouth and the muzzle are four very elastic barbs or wattles like worms. These wattles are said to be used by the animal to attract small fishes on which it preys. The value of the Sturgeon family in the fisheries of Russia and other countries is well known, every part of the fish being turned to

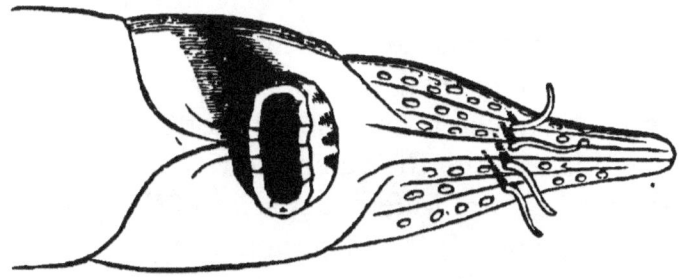

MOUTH OF STURGEON.

use. In England there are only two species. With their long snouts they poke up the bottoms of rivers to feed on the mollusks, mud-fish, insects, &c., which are buried in or frequent these spots.

The Common Sturgeon (*Accipenser sturio*) is usually from 6 to 8 feet long, but sometimes much longer, in which case they will probably weigh at least 200 lb.

This fish is migratory, but does not remain long in the sea, and is said seldom to have been taken there.

The Broad-nosed Sturgeon (*Accipenser huso*) is sometimes very large. There is a specimen of a stuffed fish in the British Museum which is 9 feet long and 5 feet 8 inches in girth. But the Sturgeons found in the great rivers Volga, Don, and Danube are said sometimes to exceed 1,000 lb. in weight; and the Russian rivers contain other species of the genus, the flesh of which is more delicate and *recherché*

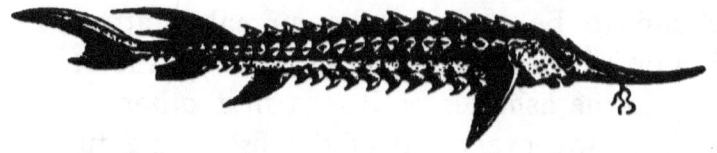

STURGEON.

than that of the common Sturgeon. The Sturgeon feeds much on small fish, and even when at sea on herrings, mackerel and small codfish. The great delicacy in Russia, the celebrated *Caviare*, is made of the roe or eggs of different species of Sturgeon. The general colour of the common Sturgeon is yellow, with a white belly.

The last family of the cartilaginous fishes is that of the Lampreys (*Petromyzidæ*). The river Lamprey (*P. fluviatilis*) is about 15 inches long; it has a round head. Its body is round for about two-thirds of its length, and then it is compressed or flattened to the end of the tail. The Lamprey was formerly much esteemed, and a surfeit of its flesh caused, history tells us, the death of Henry I.

The Sea Lamprey (*P. marinus*) goes up rivers to spawn, and remains from early spring till August, and it is mostly when in the river that it is taken. Its suctorial mouth is said to be used to draw away stones in order to have a hole or pit in which to lay its eggs. Its power in this respect is very great. The Lamprey belongs to the second order of fishes, the *Cyclostomata*, or round mouthed. The fishes of this order are in reference to their skeleton the lowest among vertebrate animals. They are tough with no scales, have a mottled skin and have no pectoral or ventral fins. A long fleshy fold of skin runs round the hind part of the body, but, as it has no rays, it can scarcely be called a fin. The mouth is dotted with small hook-like teeth. Instead of gills, the breathing organs, the Lamprey has seven little sacs or bags near the head, and each sac communicates with the water by a separate opening. Professor Owen says :—" When the Lamprey is firmly attached, as is commonly the case, to foreign bodies by means of its suctorial mouth, it is obvious that no water can pass by that aperture from the pharynx to the gills ; it is therefore alternately received and expelled by the external aperture." The water, besides gaining admission by these seven canals to the branchial sacs, is let in by the mouth if the animal so wills it, or by a round hole observable on the top of the head. This aperture has a direct communication with the pharynx. Enormous quantities of the Lamprey used to be taken in the Thames and sold to the Dutch fishermen for bait, but by draining them off in such numbers they became very scarce.

The Salmonidæ (genus *Salmo*), the most important of British river fishes, include many species. The chief of these is the king of river fishes, the Salmon (*Salmo salar*). This fish is known in its various stages by many names. Mr. Buckland considers the

Salmon to be "a sea-fish proper," but as it spends so much of its existence in fresh water, and always ascends rivers to breed, it must be considered as much of a fresh as of a salt-water fish. They deposit their eggs in the winter. When the rivers are flooded and the weirs in the rivers are covered with water, the Salmon ascend, for the quantity of water enables them to swim over these and other obstacles. The rush of water purifies the stream, and makes it more suitable for the habitat of these large fish. "The young

THE SALMON.

come out of the egg with their food provided for them in their umbilical vesicle." When this store is exhausted, they are old enough to supply themselves with food, and then feed chiefly on insects. The young salmon—in the stage called Smolts[1]—in the months of May or June, pass down the river to the sea, where they remain for some months, perhaps even a year; they then ascend the river as grilse to deposit their eggs. "A female salmon," Mr. Buckland says, "carries about 900 eggs to a pound of her weight." A heavy fish might have 30,000 eggs. Were it not for the enormous fecundity of fish many species would

[1] The stages of the Salmon are parr, smolt, grilse, and salmon.

soon become quite extinct, so many enemies and opposing influences have they. The fishes of this family are beautiful in shape and colour; they are covered with scales, the back being dark sea-green, and the belly and sides silvery-white, with black spots on the head, and irregular brown marks on the sides. These colours are only found on the full-grown Salmon. They have two dorsals, the first with soft rays, followed by a second which is smaller, formed without rays and adipose, or with a layer of fat under the skin. The body is long, the muzzle roundish, more so in the male fish. The upper jaw has a cavity into which the lower jaw fits when the mouth is closed. The young Salmon changes its colour several times before it is the full-grown Salmon. The young Salmon is grayish, striped with black. At the end of a year it has acquired a metallic hue. The other parts, according to Mr. Blanchard, are of a dazzling steel-blue; eight or ten large spots cover it as with a silvery mantle on the sides, between these spots a reddish or rather brightish-rusty iron-colour prevails; a black spot is usually observable in the middle of the operculum. The belly is of a fine diaphanous blue in the parr, the name by which the Salmon of a year old is known. The infant fry, of about half an inch in length, are very unlike what they afterwards become. After a time the parr becomes a smolt, and has a covering of silvery scales. While they are in the state of parr, they are unable to bear the salt water, but when they become smolts, they agree well with the salt water and increase rapidly in size. For a time they seem to be lost in the ocean, but at length they return to their birthplace, and are known as grilse. After depositing their eggs, the grilse return to the sea, and in a few months go back again to the river as full-grown Salmon. The male and female prepare a place for the eggs, in which they are deposited; when the male fertilizes them by shedding a milky fluid over them, after which he covers them with sand. The kelts or spent salmon, which are exhausted by laying their eggs, return to the sea to recruit, but they are weak and helpless, and many of them die on the way to the sea. Salmon make their nests and lay their eggs mostly in December and January, the darkest months, so that they are protected from many enemies by the short days and long nights of the dead of winter.

The Parr (*Salmo salmulus*) is a much disputed species. Yarrell calls it distinct from the Salmon, while other authorities say it is only a variety of the young salmon or a parr-marked trout; such is Mr. Buckland's opinion. The species or varieties, whichever they may be, in migratory Salmonidæ, are very difficult to distinguish; this is admitted by the best authorities on the subject.

The Sea Trout and Bull Trout (*S. trutta* and *S. eriox*) Mr. Buckland says are distinct species; the flesh of the Sea Trout is red and the taste is savoury when boiled, while that of the Bull Trout is white and insipid to the taste; this seems very conclusive as to a distinction in species. In some rivers the above species are increasing to the detriment of the true Salmon. The antagonism between various kinds of fish, as between various kinds of animals, is useful where the balance of nature has to be maintained; as it serves to keep each within its own limits. But should the balance be disturbed either from the interference of man, or any decrease in the natural food by its destruction in unusual weather, or any other cause, the inferior race may get the predominance, and the most valuable may be thus made to give way, to be replaced by the inferior and perhaps almost worthless. Mr. Buckland seems to fear that in some rivers the Bull Trout and other inferior kinds are getting the predominance, even so far as to drive away the true Salmon.

Salmo fario, the common Trout, is subject to considerable variety in colour; generally it is of an olive-green on the upper parts, paler on the sides; and the under parts from the mouth to the tail are of a clear

THE TROUT.

yellow. The back of the dorsal fin and the opercules are spotted with black, the flanks are ornamented with round spots of orange-red, surrounded by a circle of a pale bluish colour; the lower fins and tail are frequently yellowish, edged with black. But these colours differ according to age and locality. The Trout is covered with oblong striated scales. The head is thick, the muzzle wide and obtuse, and the eye is large. The

Trout is a greedy fish, devouring quantities of worms, small fish and fry; but its favourites are small aquatic insects. Its frequent habit of coming to the surface of the water to seize on ephemeræ has, perhaps, suggested the art of fly-fishing, which is pre-eminently followed with regard to this fish. Trout are caught in hand-nets, sieves, baskets and snares of a variety of shapes. The body of the Trout is proportionately rather long and laterally compressed. In that interesting and valuable journal, *Land and Water*, there are many instances recorded of unusually large trout; of these the one called "Lady Rodney's Trout" weighed 16 lb., and some are mentioned of even greater weight. Trout are said to live to a great age; a correspondent in the above-named paper, mentions one which had lived twenty-four years in a spring-well, without artificial food : it was sent to Mr. Buckland for his museum. Trout are found in most of the clear streams and rivers in the country. Some lakes are celebrated for them; there is the great lake Trout (*S. ferox*) and *S. cæcifer* or *levensis* (the Loch Leven Trout). The *ferox* is said to be comparable in size and weight to the Salmon. There are two varieties of it. It abounds in the lakes of Sutherlandshire and the north of Scotland, also in Lough Neagh, in Ireland. Loch Leven has long been celebrated for its trout, which in some respects approximate to salmon.

The Gillaroo Trout (*S. stomachicus*) is another species, by some writers considered only a variety of the common trout. It is taken in Lough Derg in Ireland, and in Scotland in a small lake named Mulach Corry, in the county of Sutherland. This species has a curious peculiarity in the stomach, which enables it to digest hard substances; the stomach is more capacious in proportion than that of other species, and the coat is much thicker; but, according to the anatomist Hunter, who specially examined it, and Mr. Buckland, there is no true gizzard, as had been alleged, and no essential difference in the formation of the stomach, but the wide mouth and round form of the fish give it a power of swallowing shell-fish and gravel stones, which act in assisting the digestion and may have given rise to the assertion that the fish had a gizzard.

THE SMELT.

The Smelt or Sparling (*Osmerus eperlanus*), a small fish which frequents the mouths of rivers, is of a peculiar silvery appearance, and has the smell of a cucumber; it is much esteemed in Scotland. The skin of this little fish, which is about 7 or 8 inches long, is very transparent, so that the circulation of its blood can even be perceived when examined through a microscope. The tail is forked, and the scales are so loosely attached that they easily drop off. The Hebridal Smelt is another species mentioned by writers. The Grayling (*Thymallus umber*, or *Coregonus thymallus*) is a fish as beautiful to look upon as it is good for the table. It seldom weighs more than 2 lb. "It is remarkable for the height of the first dorsal fin, which is crossed with square dusky spots; the head and back are dusky-brown, the sides of the body of a light yellowish-brown, with golden-

THE GRAYLING.

green and blue reflections, and about fifteen longitudinal lines with scattered black spots. The head is small and the scales large. It varies from 9 to 14 inches in length."

When freshly taken from the water it has a pleasant aromatic smell, owing probably to the water-thyme and other sweet-smelling plants it feeds upon.

THE GWYNIAD.

In some of the lakes of Ireland a fish of this family called the Pollan (*Coregonus pollan*) is abundant, and the Gwyniad (*C. fera*), another species, is found in the lakes of Cumberland and Wales. In the heat of summer it seeks shelter in the deepest parts of the lakes. This fish is not caught with a rod but with a sweeping net. It is a small fish, seldom exceeding 10 inches. The body is compressed and covered with moderate-sized scales; the eye is large, the mouth wide, and the under jaw projects; small teeth are placed on the tongue. The back is brown and the sides yellowish, the cheeks are white, and there are pale yellow lines along the body.

The Pollan is called the fresh-water herring from its resemblance to that fish, and from the immense shoals in which it is taken. It is almost exclusively taken in the large lakes of Ireland, though sometimes found in rivers. They spawn in November and December on the bottom of the lake. It is from 9 to 12 or 13 inches in length. The Vendace (*C. Marænula*) is found—Mr. Buckland says—exclusively in the lochs of Dumfriesshire. This fish is considered a great delicacy, and is a good deal like the Smelt in flavour. Its food is chiefly small animalcules; it is taken solely in nets and not by anglers. It is said to have been imported from France by Mary Queen of Scots. The Argentine (*Scopelus Humboldtii*) is so scarce that it may be doubted whether it should be enumerated among British fishes. Mr. Buckland says he has never seen a specimen.

CHAPTER VII.

SOME OF THE TYPICAL RIVERS AND THEIR FISHERIES.

THE Tyne, which is formed of two streams meeting near Hexham in Northumberland, has thence a course of thirty-five miles, till it reaches the German Ocean. From the source of its northern stream the distance is about eighty miles; this stream rises in the Cheviot Hills; the southern branch rises in the mountains of Cumberland. The Carlisle and Newcastle Railway passes along the course of this branch till it reaches the latter city. The river is navigable for vessels of four hundred tons up to Newcastle, and for several miles further up by boats. The Tyne's greatest importance is owing to its carrying to the sea the coals for which the district has so long been famous. In former years the Tyne salmon were abundant, but the locks built on the river, and the pollution caused by traffic and manufactories have worked their usual havoc. In prospect of Parliament interfering to protect the fishery and thus to restore former prosperity, the celebrated Thomas Bewick, "father of wood-engraving," wrote the following letter :—

"NEWCASTLE, *April 26th,* 1824.

" I have met with few things in passing through life that have given me more pleasure than the information you have this morning imparted to me respecting

Mr. Brandling's intentions of laying before Parliament the various causes which, taken together, throw obstructions in the way of the salmon-tribe breeding in the Tyne in the same overflowing numbers as of old; and in putting together a few remarks in as short a way as time permits, to state my opinion as to the reasons for such an immense falling-off. When a boy, from about the year 1760 to 1767, I was frequently sent by my parents to the fishermen at Eltringham Ford to purchase a salmon. I was always desired not to pay twopence per pound, and I commonly paid only a penny, and sometimes three halfpence, before or perhaps about this time. I have been told that an article had been always inserted in every indenture of apprenticeship in Newcastle, that the apprentices were not to be forced to eat salmon above twice a week, and the same bargain was made with common servants. I hope the time will shortly come when the same overflowing bounty of providence will again enrich my beloved Tyne. Whatever obstructions are thrown in the way to prevent the salmon from ascending so far up every river and rivulet as they can reach for the purpose of spawning, is the first and great cause of the breed being thinned; therefore, every weir and every dam ought to be removed. The fishermen's weirs are bad, but of these Bywell is the worst; they both have their rise in a greedy and selfish disposition to prevent other fisheries from partaking in a due share of the fish. You will be able as well as I can to point out to Mr. Brandling the evils arising from the uses of nets of various kinds, which obstruct the fish from ascending the river, as well as those which arrest the fry on their way to the sea. Should the business of weirs and dams be settled, then river conservators should be appointed to guard the spawning fish (usually called kepper fish) from being killed while they are in this sickly state. These conservators ought also to be empowered totally to prevent the wicked destruction occasioned by putting lime into burns, as they kill millions of spawn, as well as every living creature in the water within its extended reach. Fishermen may grudge to see the fair angler fill his creel with a few scores of the fry which would perhaps return to them as salmon in a short time; but when it is considered that a pair of salmon will breed more of this fry than all the fair anglers can catch from the head to the foot of the river in a season, I think it is cruel to debar such from enjoying this diversion. Fish are not like game, which are fed by the farmer; their food costs nobody anything, and ought only to be preserved so far as they may be for the public good; therefore, I have always felt disgusted at what is called preserved rivers. In these, because they run through the land of some freeholder, the fish are usually claimed as his own;—the disposition which dictated claims of this kind is the same which would, if it could, sell the use of the sun and the rain;—where the angler is debarred the most delightful of all recreations, which ought to be his birthright, particularly of the sedentary and studious; it is the healthiest and most innocent of all diversions, it unbends the mind and enables such and the pale artist to return to his avocations or studies with renovated energy, to labour for his own and the public good.

I ought also to name to you the uncommon destruction of the fry which frequently happens when they are hastening to the sea by the stopping of a mill-race with thorns, and then letting the water off some way by which it has been known that a cartload of fry have been taken at once.

"I named to you the kind of weirs which ought to be made out of tide-mark, to increase the depth of every river in the middle, by which a more equal chance would be given to all fishermen to come in for their due share, and at much less trouble and expense than they have hitherto been put to, and would besides open a free passage for the fish whenever they instinctively ascend to the proper spawning-ground.

"(Signed) THOMAS BEWICK."

A good many years passed before much benefit was done in the way of protection to the fisheries on the Tyne, but when, about twelve years ago, energetic measures were set on foot, the advantages were soon apparent. None could fish without a license. In 1867 salmon became more plentiful than they had been known for generations. A writer in *Land and Water* for that year says, that we may fairly say the produce of the river was well over ten thousand fish. It is stated that some of the salmon weighed as much as from 30 to 40 lb. The fish when not full-grown is called grilse; it is not equal in flavour to the full-grown fish. The same writer further on in the season says :—"The fishery continues to be very good indeed; I heard of one which was sold at Gateshead weighing 43 lb. One was caught of a most singular shape, it was barely two feet long, yet it weighed 20 lb."

But heavy as 40 lb. weight may appear for a salmon, it is said that forty years ago, they were taken as heavy as 75 lb. This writer of 1867, says that "the river was positively alive with salmon, making their way up. The other day people were standing all up the quay-side and on the bridge at Newcastle watching them. Near Hawks' works some timber was moored and a couple of salmon were seen to jump clean out of the water and fall upon the raft; they were soon captured. Three tons weight were said to have been taken one day near Lemington, four miles above Newcastle."

There are at least three rivers called Derwent in England, but the most important and interesting is that which rises in Borrowdale, in the county of Cumberland, and flows into the lake known as the Derwentwater, and thence into the Bassenthwaite Water, from whence it issues, and passing Cockermouth enters the Irish Sea at Workington. The Greta, a small river which runs into the Derwent, is yet of sufficient volume after heavy rains, when floods are high, to turn back the river Derwent, and cause a great accumulation in the lake. The four miles' distance between Derwentwater and Bassenthwaite is a famous fishing-ground, protected during the season by the Angling Association. This is a part of the country equalling in romantic beauty and celebrity any part of our favoured isle. Derwentwater is about a mile and a half across its widest part, and is about four miles long. There are several wooded islets in it, besides a mass of floating soft land. Its high rocky banks, with the contrast of the wooded islands, make one of the loveliest pictures to be seen in England. The lake Bassenthwaite, though pretty, does not equal Derwentwater; the former is not quite so long—it is 225 feet higher than the sea. Another river, called the Cocker, also runs into the Derwent, besides smaller streams.

In the eighteenth century the river Derwent was famed for its salmon fisheries. Opposing influences were always at work to check the productiveness of the best fishing streams; but it is in the present century, when modern mills and manufactories have come to their height, that the noise and pollution of the rivers, with their living inhabitants, have too often changed them from streams of wealth and purity to dull muddiness and unhealthy animal life. Where many tourists go, and the human natives of the district lose their simplicity, there too the fish are apt to become scarce; and if mills or dye-works should be erected, the finny folk "cannot bear the muddy stream or brook where manufacture grimly grinds its iron teeth, for it pollutes earth and poisons life with smoke."[1] The amazing cheapness of salmon in former years was due doubtless to various causes. Money was more valuable and went much further in the last century and first fifty years of the present than it does now. The great increase of population causes perhaps a manifold greater demand for food in the present day, and when we add to this the ravages

---

[1] 'Book of Nature and Book of Man,' p. 169.

of the enemies of our fish, it is not difficult to discern the cause of variation in price. We read in the account in *Land and Water* (March, 1866), of the salmon fishery of the Derwent, that salmon at Workington, at the beginning of the present century, might be had *from one penny to sixpence per pound;* fifty years later its cost was from sixpence to one shilling per pound.

The river was to a certain extent protected as long ago as the year 1285. Sixty or seventy years ago the celebrated Hall fisheries of the Derwent were let out at £120 a year. The tenant was able to make the immense profit of £10,400 per annum. But twenty years ago the tenant, although only paying £72 a year, and salmon worth double or much more than it is now, yet such was the scarcity of fish, that small as the rent appears there was little more sold than to pay for it and the expense of working.

The method of fishing at this period was fatal to the increase of fish. A weir was built in a sloping form entirely across the river, and over the weir was placed a row of perpendicular iron bars; between the iron bars coops or fish-traps were placed. The salmon, when passing up the river, were caught in these traps; but in order more effectually to catch every fish, as it was hoped, at a later period, the iron bars were altered in position and placed horizontally, also much more closely. The weir, the bars, and the coops were ruinous to the fish, and except when floods came heavily down the river, few, if any, could pass the obstruction on their way up. However, in 1804, an Act of Parliament, called the "Solway Act," fixed the "fence" time from the 25th of September to the 10th of March, and ordered that the bars of iron should be placed not closer together than 3 inches. But a further Act of Parliament in 1809 gave the lessee of Lord Lonsdale's fishery, which only extended for two miles and a half of the river, the right to fish from the 10th of February to the 10th of October. Thus the owners of all the length of the river beyond Lonsdale's property, and which extended for thirty-three miles, had no protection for their fish, and this contributed gradually to destroy the produce of the river, which had been so wonderfully prolific.

Matters went on in this way till the year 1861, when a new Act of Parliament for English river fisheries altered the face of affairs. Near the mouth of the Derwent the Scotch method of spearing salmon still further contributed to its well-nigh extermination. One method was for a man mounted on horseback to follow the swimming fish in the water up to the horse's belly, holding the bridle in one hand and galloping after the fish; when sufficiently near he struck the fish with a spear, which he had in both hands. He seldom, if skilful, missed striking the fish. Then raising the transfixed fish to the surface of the water, he turned round the horse's head to the shore and ran the salmon on dry land before he dismounted. In this manner forty or fifty salmon were said to be killed in one day. Another method of catching the fish was by stake-nets, erected near the sea at Workington.

It is stated in the account of river fisheries in *Land and Water* for March 31, 1866, that as many as eleven cartloads of fish were sometimes taken in one day; but by the Act of Parliament for 1861 the use of these stakes was prohibited. "However, in defiance of this statute, they were continued to a certain extent here, Lord Lonsdale claiming a right to erect them on the foreshore of his manor of Seaton; and in 1865, at the Spring Assizes in Carlisle, the river was emancipated by the Cockermouth Association, who, through their secretary, challenged the right of Lord Lonsdale's tenant to erect a stake-net near the mouth of the Derwent. The challenge was given by cutting down part of the net. A verdict was obtained against the tenant, so that this harm came to an end."

Among the enemies of the fish must be mentioned the poachers, who are, or were some years ago, scattered along the river-side. They used to subscribe among themselves to pay the cost of any fines which might be imposed on them when caught. So much has been done by legislation and by fish-culture of late years that there is much hope that in time this beautiful river may return to its former wonderful power in supporting a great abundance of fine fish.

The town of Workington is famous for being the landing-place of the unfortunate Mary Queen of Scots when flying from Scotland. Miss Strickland says Mary's mind misgave her when fairly out at sea and under sail for England, and she said she would go to France. The boatmen made an ineffectual attempt to change their course, but the wind and tide were contrary, and carried the little vessel rapidly across the Firth of Solway and into the harbour of Workington. The voyage is said to have been performed in four hours. Sixteen persons accompanied the Queen. As it was Sunday evening, the general holiday of high and low, an unusual number of people assembled to see the Scotch boat come in. Rude as this vessel was, she excited lively curiosity, for it was instantly perceived that her passengers were neither fisher folk, colliers, nor Kirkcudbright traders. It needed not regal ornaments or robes of purple to proclaim Mary's rank, exhausted with grief and fatigue though she had been for the last three nights, and wearing the travel-soiled garments of white silk in which she had fled from the lost battle of Langside. The moment she stepped on shore she was recognized as the fugitive Queen of Scotland.

The river Dee is one of the most interesting streams of the north-west of England and Wales. Its source is Bala Lake, Merionethshire. It flows through the beautiful vale of Llangollen and Wynnstay, and then running north separates Denbigh from Flintshire and Cheshire, and runs into the Irish Sea; it is, therefore, more a Welsh than an English river.

A good deal of fine fish comes from this river, though not so celebrated in this way as its Scotch namesake. Many traps, weirs, and obstacles of various kinds oppose the passage, up and down, of the noble salmon, yet it has been able to reach the upper waters in its yearly progress from the sea, and still is tolerably abundant, notwithstanding poachers and more systematic commercial enemies. Fine trout and other mountain-stream fish are found in this river. But in early times the Dee was most abundantly stocked with salmon. In Domesday Book, at the time of the Norman conquest, when a survey of the fisheries of the country was taken, there were "established fisheries on this river, some belonging to religious houses, and others to private manors. There was one fishery at this time some miles above Chester, employing six fishermen and yielding one thousand salmon, but this became exhausted."

Other fisheries were on the Dee at this period. St. Giles's hospital for lepers was founded and given "three stalls in Dee, under the seal of Chester Exchequer." The privileges of this hospital were confirmed by Hugh Kevelior and Edward III. Among these privileges they claimed "one salmon from every horseload or cartload of salmon brought to the Chester market, and to have also one boat with a fisherman, above or below Dee bridge, with stallnette, flotenette, or dragnette, or any other kind of nette, night and day, and three stalls in Dee, called single-line stalls." And the old Chester Cathedral, the Abbey of St. Werbergh's, was found at the time of making the Domesday survey to possess fisheries in Dee; and when the Conqueror gave to Hugh Lupus the earldom of Chester, Hugh, by charter, bestowed upon the monks of St. Werbergh's, over and above their own fisheries, a tenth of all fish taken in the Dee. And Hugh Lupus gave fisheries to others as to the monks.

A law was made, locally called the "Ould custom and law of Chester, to the effect that the stall-netts in the water of Dee were not to exceed the size of nine fathom, and the dog-rope one fathom; and if the stakes whereunto the dog-rope is attached be higher in the water than to a man's knee, then the said netts be forfaited." The size of the boats and length of oars on the river was not to exceed 20 feet for the boats, and 16 for the oars, so that the boat might pass up the centre of the river without coming in contact with the staves and nets on the banks and sides of the river.

There was an ancient office called the "Sergeancy of the Dee," by which a man was appointed to see that these rules were carried out. This dates as far back as the Norman conquest. This office was hereditary. In the reign of Edward II., Robert Grosvenor de Eton claimed this right, which had been neglected in his family for some generations. "The Sergeancy of the Dee, from Eton Weir to Arnoldshyre (a rock opposite Chester, now called Arnold's eye), by the service of clearing the rivers from all nets improperly placed there; and to have a moiety of all nets forfeited and of all the fish therein as far as stall-nets are placed, viz., from Dee Bridge to Blakene, thence to Arnoldshyre, to have one out of all the nets taken, and to have a ferry boat at Eton over the water . . . and to have toll from every flotte at Eaton passing through his weir; and to have waifs and wrecks on his manor of Eton, and two stall-nets and two free boats in Dee."

These custodians of the river had not fulfilled their offices very faithfully in early days, for even in the thirteenth century there was a great falling-off in the yield of fish. And in the days of Edward I., a Dee salmon, according to the ancient records of Chester, was worth ten shillings; in those days this was so large a sum that one is tempted to think there must be some error. In this reign it is recorded that, "On the morrow after the day of Palms (we suppose Palm Sunday), a certain William, son of Aldos, and others were brought before the Magistrate and fined for unlawfully fishing below the bridge of Chester, and catching twenty salmon worth twenty marks (6s. 8d.), and shortly after fished again and caught a salmon worth ten shillings and more. On the same day, others were fined for catching salmon worth one mark or more." Other laws were passed, at different periods, to protect the fishery.

The office of Sergeancy is claimed by the Duke of Westminster, as still belonging to the Manor of Eaton. The river is so much frequented by salmon and trout anglers, that the inns on the banks of the river can at times scarcely afford accommodation for them.

The derivation of the name Dee has occasioned much disputation. Some say it comes from Ddu (black), others from Dwy (two), because the river is said to come in its very beginning from two small streamlets; but Pennant, a great authority, denies this, and says Dee is derived from Duw, signifying divine, because to the river was attributed sacred powers, and that it was well known that the Welsh were fond of giving divine honours to their rivers and fountains.

Giraldus Cambrensis, who travelled through Wales in 1188, says that, up to his time, prophetic properties were attributed to the Dee :—

"Dée, which Britons long ygone
Did call divine, that doth by Chester tend."

And Drayton says :—

"And lastly, Holy Dee, whose prayers were highly prized
As one in heavenly things devoutly exercised,
Who changing of his foods by divination had
Foretold the neighbouring folks of good or bad ;
In their intended course sith needs they will proceed,
His benediction sends in way of happy speed."

Milton, alluding to the supposed supernatural powers of the Dee, says :—

"Where Deva spreads her wizard stream."

The Dee, which runs through the largest lake in Wales, Llyn Tegid, or Bala, is 137 feet above the level of the sea. The lake in some parts is 138 feet deep, of clear water, besides a bed of mud. The fish called the Gwyniad frequents the deep parts of the lake. This fish somewhat resembles the whiting, and is much prized for its delicate flavour. The salmon and trout prefer running water, and if they enter lakes through the streams in which they swim, they generally prefer to pass on. The gwyniad is a lake fish, feeding chiefly on shell-fish and water-plants. The *Lobelia Dortmanna*, growing in the shallows, is a favourite food, and is said to give a peculiarly agreeable flavour to the fish. The pike is also abundant in this lake, as are the tench, roach, perch, and eels. The pike, says Tennant, has been caught here twenty-five pounds in weight, trout twenty-two pounds, perch ten pounds, and gwyniad five pounds. But fish of these weights, says a writer in *Land and Water*,[1] in 1866, are seldom, if ever, met with in these days. In the old stories about this lake it was said that, let the

Lobelia Dortmanna.

rain be never so abundant, or the summer's heat never so parching, the quantity of water neither increased nor decreased; wind alone caused any of the water to

[1] Mark Heron.

swell over the banks. The water-nymph of poetry rivalled the fays of the mountains, and each praised their own domain. The nymphs rose from their sedgy bowers below, and rudely interrupted the Merioneth mountains, which had been shouting their praises with such vehemence that the sounds reached to the Courts of Neptune far off in the depths of the sea, and awoke him—

> "Who, full of dread,
> Thrice threw his three-forked mace about his grizly head,
> And thrice above the rocks his forehead raised to see
> Amongst the high-topt hills what tumult it could be;
> So that with very sweat Cadoridric[1] did drop,
> And mighty Rauran[2] shook his proud sky-kissing top."

The Llyn water-nymphs could bear it no longer—

> "With brows besmeared with ooze, their locks with dew besprent,"

they came from their watery home uttering a "suddaine fearful noise," took the mountains by surprise, and put them to silence. "Old Snowdon from the neighbouring shire came to the rescue of his brother hills of Merioneth, and before the sound of his venerated voice even the proud and daring water-nymphs gave way, and let the mountains speak." But the most wonderful of all doings of great lake and enchanted river is the fable that the Dee passes, or did once pass, through the lake an undivided stream, and issued the same pure unmixed water as it entered. This fanciful idea concerning favourite rivers finds a parallel in the notion that the Rhone passes through

---

[1] Cader Idris. [2] Arran.

the Lake of Geneva unmixed and uncontaminated. In Homer there is the same idea respecting Titaresius. Pope says—

> "The pleasing Titaresius glides,
> And into Peneus rolls his easy tides;
> Yet o'er the silver surface pure they flow,
> The sacred stream unmixed with streams below.
> Sacred and awful! from the dark abodes
> Styx pours them forth, the awful oath of gods."

To call forth the embodiment and personification of the grand objects in Nature, one must inhabit, at least in thought, those regions where rivers, lakes, mountains, and grand sky-sights abound; for where can be met with in poetry the dull plain, the insignificant down even, or the commercial dwellings where man's works are the chief objects in view, not those of the Creator; and human contrivance fails to inspire grand conceptions? We need great objects far above us to suggest great ideas.

The Eden is a river from whence comes much good salmon, but like all the other rivers of our country, before the revival of fish protection, and attention to the state of the fisheries, much diminution in the supply had occurred. In 1866 a correspondent of *Land and Water* writes:—"Mr. Parker caught with the casting-rod five fine salmon on Saturday and five on Monday. I caught one yesterday with the casting-rod, weighing twenty-five pounds. The river Eden is swarming with salmon." This fine fish throws all other river fish, at least in England, into the shade.

The Eden meets the Calder and Petrie rivers near the city of Carlisle, and flows into the Solway Firth;

it rises in the mountains of Westmoreland; it is beyond the "Lake District," and therefore has not the intense interest for the tourist in these parts which the Derwent has. In March, 1879, owners of salmon-fisheries were much disturbed by the alarming reports of a fungoid disease on the salmon of the Eden. It generally first attacked the head, and extended from thence to the tail. Ulcerating sores ensued, and the fish was generally blinded by the spread of the fungus in the eyes. Maddened by the irritation thus caused, and not being able to see its way, the fish generally killed itself by coming violently against some obstacle. Mr. Frank Buckland, Inspector of Salmon Fisheries, said it was due to overcrowding in the river Eden. The fungus was very contagious and malignant, irritating the skin of the operators who dissected the salmon. The police collected and buried 1,271 salmon affected by the disease. Mr. Frank Buckland, with a view to his seventeenth annual Report laid before Parliament, in 1878, asked me to examine the fungus on the Eden fish, and the result of my examination being printed in the Report, I quote it here :—

"The growth and transformation of the microfungi are amongst the most remarkable of organic metamorphoses. Mr. Frank Buckland having placed in my hands a piece of salmon's fin infected with fungi, which he cut off from a fish in my presence, I am able to say, from having carefully examined it microscopically, that it is the *Saprolegnia ferox*. Mr. Frank Buckland dissected in my presence the stomach of a salmon, and found in it some granules which he placed under the microscope. On examining the figures in Cornu's monograph, I am convinced that they represent a stage of development of the *Saprolegnia*.

"Mons. Max Cornu, in his 'Monographe des Saprolegniées' in *Annales des Sciences Naturelles*, 1872, says, the family of the *Saprolegnia* constitute a natural group of aquatic fungi, to which species have been added from time to time. The principal writers on the subject have been German, such as Schlieden, Unger, Braun, Pringsheim, de Bary; and M. Muret, a Frenchman, has also written briefly on the subject, and has given good and accurate drawings which better illustrate the subject than lengthy verbal description. But according to MM. de Brongniart and J. Decaisne no general work has yet been published, for the above-mentioned authors have only written short comments on these fungi, which are scattered throughout various periodicals. The difficulty of studying these genera is great, for they can only properly be observed when in a living state. The rapidity of their growth, their short lives, and rapid decay make the procuring of them in the proper state a difficult matter. M. Brongniart says he met with some of these species which had been previously described; he himself discovered others not described, and felt convinced that many yet remained still unknown. He considered the manner of reproduction and method of propagation to be the most important points for observation. Learned men still differ much on these points. M. Cornu says the germs may be disseminated either through the air or through water; thus showing that the vehicle of distribution is unimportant, the plasmatic (*plasmatique*) part is alone essential.

"In examining the reproductive organs of *Saprolegnia ferox* at first only the (oogones) eggs and spores are visible. The spores are cells sometimes of an irregular cylindrical shape, at others of an oviform shape, as in *Saprolegnia achyla*, but in some cases variable and in others uniform. As at first sight the male organs appear wanting, M. Cornu was led in his researches to inquire whether the spores being variable in form might not really represent in some instances the male organs. In the genus *Monoblepharis* the sexual reproduction has been observed, he states, in two species having cylindrical spores. In some small special spores, more or less isolated from accompanying spores, have been found antherozoids born or produced in the spores. They have the form of zoospores, and are capable of fecundating the gonospheria. These two species of *Saprolegnia*, he states, are the same in their fructification if we only consider the vegetable filaments and the spores, but the apparatus for reproduction presents differences.

"In one species the egg is spherical at the end of the filament, or it may be abnormally-shaped laterally, in consequence of the encroachment of the axis under it. This form of egg is characteristic and sufficient to distinguish it from the other: he calls it *Monoblepharis sphærica*.

"The extremity of the filament which is to be changed into the egg is enclosed in the mass, then swells into a spherical form; the plasma collects at this extremity, and soon condenses into a quantity of oleaginous grains, yellow and shining; during this time the swelling sphere is separated and closes up. Afterwards the egg becomes more consolidated.

"Immediately underneath it there is a place in which the plasma in small grains, irregular and confused, presents a different aspect, and in fact shows a spore in course of formation. Its length is now from four to six times its breadth. It is not dilated, and is in fact but a continuation of the filament which supports it. A prolongation of the filament only can be seen directed towards the egg, as if the axis grew laterally, but the development goes no farther, and this part is isolated by closing itself where it adheres to the filament; it thus presents all the characteristics of a little sphere. It is the *anthéridie*. Like the egg (oogone) it is always of the same form. Therefore the eggs of *M. sphærica* are spherical, and furnished with a single anther situated below them in the filament. They are generally solitary, and rarely double. These organs are the same in all individuals. But in the other species this is far from being the case. The shape of the eggs is different, whether they bear anthers (*anthéridie*) or not, whether they are grouped together or otherwise; the grouping besides is different. M. Cornu proposes to give them the name of *M. polymorphia*.

"When the egg is isolated it is of an oval shape, larger at the upper end and more pointed at the lower end. When several eggs are fastened to one filament, this shape is a little altered; the lower has a sort of side platform on which the nearest is leaning; and so on with the rest from one to another. Perhaps a dozen may be thus placed on one another, sometimes in groups, sometimes singly.

"The form of the anthers also differs. Sometimes, as in *M. sphærica*, they support the egg, at others they separate from the lower part of the egg, or are situated over it. Sometimes these layers are mingled. Several anthers may be placed in a row like the spores which they resemble, except in shape. Several of them may be on one egg, and a group of eggs on the other hand may be entirely without anthers. In this case they are generally situated at the end of the filaments. They would be confused with the spore were it not for the diversity of shape, and above all by the presence of a great number of them on the eggs themselves, or very near them. The difference between the two species is well characterized by the appearance of the *young* eggs, but the difference is much greater during and after fecundation.

"When the fecundation is about to be accomplished, a change appears in the egg. The upper part is formed at first into a point, which afterwards opens, and the oleaginous globules in motion, first towards the opening and afterwards descending, meet the ascending antherozoids, which, breaking the plasma, ascend laterally. The fecundation takes place by the fusion of the elements of the gonosphore with those of the antherozoid after the latter has penetrated into the interior of the egg. Each antherozoid as it escapes through the plasma touches the (*oogone*) or egg. If it fails to do this, it escapes into the liquid like pure water, which is above or under the globules. If it touches the egg, it adheres to its surface. The fecundation thus is performed by the entrance of an antherozoid through the opening of the egg. The antherozoids climb on the surface of the egg, until finding the opening they enter and spread the viscous matter of which they are composed on the surface of the gonospheria.

"This occupies a few minutes. Then the gonospheria by a slow movement encloses itself in the egg and remains there about five minutes. The fecundated mass is then formed into a ball environed in a covering or skin which escapes through the opening in the original egg.

"M. Cornu has watched the whole process completed in about an hour.

"He gives many minute details which we cannot enter into, and describes the variations in the comportment of the two species he has specially watched.

"He differs considerably in his deductions from former botanists who have studied these difficult and delicate species.

"What are the causes of the ravages of this fungoid? I do not hesitate to say unwholesome water, produced by river pollution, either by sewage, factory or mine drainage, rendering it difficult for the higher organic life to live, and facilitating the development of the lower fungoid life. The wide dissemination of the fungus on the salmon both in England and on the continent forbids us assigning any special impurity in the water to its development, but clean rivers free from decomposing dead fish and chemical washings are the least liable to be affected. The water of every river where the fish are found with this disease should be carefully analysed with a view to trace the source of impurity, and means taken to destroy the diseased fish, which it should be legal to take even in close time by remitting all penalties if fish can be proved to be diseased."

The Hodder, a tributary of the Ribble, has been carefully protected for a salmon stream. The fish-hatching houses at Whitwell and higher up the river are so interesting that they would tempt many to try experiments for amusement, if not for profit. The salmon-ladders seem curious to the uninitiated; they are made to help the fish when they ascend the rivers at the breeding season. A letter in *Land and Water*, dated February 23, 1867, says, addressing Mr. Buckland, "You will remember when at Knaresborough going down to 'Scruton's Dam' to see the salmon leap. At the opposite side from where we stood is the 'Mill-wheel,' and then alongside there is a passage which can be opened or closed by a 'clow' or 'clough.' Last year about this time the river was much swollen, and through this passage the water was rushing with great force. The top of the passage is boarded over, and on my arrival there on Sunday afternoon a salmon, said to be about ten pounds weight, was lying struggling on the boards, and I felt certain this fish lived from twenty to twenty-five minutes out of the water. We spectators were all on the opposite side from the fish. It was impossible to cross or make the parties residing at the mill aware of what was going on in consequence of the noise at the dam. A lad started round by the nearest bridge, whilst we stood watching the struggling salmon, sometimes getting his tail overhanging the water again, and the next struggle landing him further on the board. The fish was just dead when the lad got round, a distance of two miles. Just before my arrival a man had secured a salmon, which had taken the same fatal leap, and from the cool manner in which he came and took the salmon those who saw him remarked that he had been at the job before."

Here a salmon-ladder was much wanted to save the fish in their desperate efforts to ascend the stream. It is to be hoped the want has been long supplied; but we quote this anecdote as illustrative of the habits of this fish, and of the dangers they are often exposed to. Those who are most experienced in salmon rearing and protection think that the more the rivers and fish can be left to a state of nature the better; but as in this country a state of nature, strictly speaking, is impossible, our streams must be protected, though it may be by bringing one artificial system to counteract another.

The Trent is a favourite river with the anglers of Nottingham, who, after a long frost, think the breaking up an excellent time to catch the pike, perch, and chub. Sometimes good jack are caught below Nottingham. The low situation of the neighbourhood makes it apt to be flooded after heavy rains. As in other of the larger rivers, salmon, which used to be plentiful in the Trent, became scarce. The grayling, a finely flavoured fish, not very abundant anywhere, is found in the Trent and Dove, which runs into it.

The Ouse joins the Trent, and forms the Humber

between Yorkshire and Lincolnshire. Many rivers of Yorkshire run into the Humber, as also some of Lincoln, Leicestershire, and Nottinghamshire. There is also the Great Ouse, which flows through Northamptonshire, Bucks, Bedford, Huntingdon, Cambridge, and Norfolk. This has a course of about 160 miles, till it reaches the Wash at King's Lynn. This river has been the subject of verse in the simple lines of Cowper about the dog, which, seeing its master anxious to secure a water-lily, swam to the spot and brought the flower to his master's feet. A flat country or one slightly undulating, if threaded by a fine stream, may inspire poetry, though not so readily as the lofty mountain and sparkling cascade. The sweet but rather prosaic lines which Cowper has addressed to the animals and birds of his neighbourhood, seem to partake of tame, homely, rural beauties, rather than of the poetically grand.

The South of Devon is so full of beautiful little watering-places that it is often difficult to say which is the most attractive. At Budleigh Salterton the little river Otter runs into the English Channel. Years ago, when I was a boy, I used to fish in this stream; a single trout, a pope, a roach, and perhaps a lot of bleak, were all I could make my own. A few salmon were in the river, but I was not lucky enough to secure even one. A few miles up the stream, near Honiton, the fishing was good. Once I was out on a fishing excursion, not far from the town; I felt a tremendous pull at my hook, placed myself firmly on the bank, gave the fish the utmost extent of the rod, and waited the result. The pulling became harder, and I felt sure I had got a twenty pound salmon at least, when out flew the top joint of the rod and was pulled violently down the stream. I watched it float with great velocity down the current and followed it for some hundred yards. It was apparently dragged by some powerful force exerted under the water. At last the rod stopped, caught among some reeds. I looked backwards and forwards, and at length saw beyond me, in an inaccessible marsh, an otter, which had just landed with a large fish in its paw; *this otter was my twenty-pound salmon.* An old fisherman came up to me, who had been more successful than I, and was returning with his basket full. I wanted a piece of string, to which I thought if I attached a stone I could throw it over the rod and drag it ashore. I had no string with me, and all my line was in the water; but the old man lent me his rod and line, and with it I got my own ashore. I found the hook and gut had been bitten off by the otter. This old man was very good-natured, and took me to one or two pools where the fish, though small, were greedy biters, and I caught them by the dozen. They were about four inches long, and of a species unknown to me. This man was full of anecdotes of grand captures, and invited me to give him an eel-hunt the next day. A small rivulet crosses the lower part of the town; this had shrunk to very narrow dimensions on account of the dryness of the weather, and the fish that were usually there had entirely disappeared. These, the old fisherman told me, had congregated in a deep subterranean pool, where they were, he said, 'as thick as herrings in a barrel.' We went accordingly to this hole, which had formed the lower story of some building, but had become filled with water and so converted into a pond. It was being pumped out, and the old fisherman had beforehand offered £10 for the fish it contained; this being thought a ridiculously high price was at once closed with. We waited until all the water was drawn off but about two feet, and then the old man waded in. The chamber was about 20 feet by 18, and about the same depth; its walls were clothed with green weed. I got upon a wall, which was just out of the water, and gazed into the basin, which appeared a perfect swarm of fish. I waded in until the water was two inches above the knees, and could hardly stand for the wriggling mass of eels. I noticed one or two savage-looking pike, and fearing they might bite me withdrew to the wall. One followed me, glaring as fiercely as a fish could, and, making a desperate spring, leaped high and dry on the top of the wall. It was a three-pound pike, and I secured it by standing on its tail. The water was still further drawn off until there were only a few inches left—one moving mass of eels, roach, loach, carp, pike, tench, and dace. The old fisherman could hardly contain himself with astonishment; he had ordered two market carts to convey the fish away, but this was not enough. He packed thirty large baskets of eels in one van, and sent them to meet the train for London. He got in all many hundredweights of fish, being the greatest capture ever made in the neighbourhood. Nine-tenths of the amount were eels, but there were about $1\frac{1}{2}$ hundredweight of carp, and as much more of other species. This had been a suspected preserve for many years, but the dry weather had caused the fish for miles to congregate in it. I returned home all slimy and a fearful figure of dirt, but immensely excited with my day's fishing, which had I not seen I would not have believed.

The Exe, on which stands the ancient and pretty city of Exeter, rises in Exmoor, a hilly part of Somersetshire. It flows into the English Channel at Exmouth, one of the beautifully-situated watering-places for which the county is so celebrated. The views from the banks above the mouth of this river, which there forms an estuary navigable to Topsham, four miles from the sea, are very beautiful though not grand or so romantic as those from the Dart. The banks of the Exe afford a great number of fine plants, among which may be mentioned the Marsh Marigold (*Calantha vulgaris*), with its magnificent golden blossoms; the Purple Loosestrife (*Lythrum salicaria*), with its richly and finely-cut purple flowers arranged in a spike; several species of Sedges (*Carex*); the Yellow Iris (*Iris Pseud-Acorus*), the hairy Willow Herb (*Epilobium hirsutum*), and the Water Star-wort (*Callitriche verna*). In a wood a little away from the bank of the Exe, about three miles from Exeter, in the month of May, the author, with a few friends, passed a day. "It was a fine season and vegetation displayed the glowing precocity of a youth of genius. We alighted on the greensward and entered a copse; the Harebell (*Hyacinthus non scriptus*), the Primrose (*Primula vulgaris*), the early Purple Orchis (*O. mascula*), and the Cuckoo Pint (*Arum maculatum*) burst on our eyes in such luxuriant profusion that they appeared to have been gathered by kind Nature from a large area. We separated, taking different directions in the little wood. It was not to me unexplored territory, but I had not seen it at this sweet season of the year. I passed a clear brook, over which hovered swarms of gnats; on my right hand was a hawthorn hedge in full blossom, and at every breath of wind a shower of blood-stained snow fell on the mossy bank beneath. A few yards further on the trees were very dense, and there I found a stone on which I sat to count my flowers. I heard a bubbling further on, which led me by its gentle voice to a little pool about 6 feet wide, fed by a small stream of water. It was shaded by a fine oak, and a natural wall of rock, about 10 feet high, bounded half its extent. The rocks were rugged, but from their surface a luxuriant crop of ferns and mosses grew; shaded from the sun they displayed those delicate green tints which live not in the bright light. The water was cool and refreshing, and as it poured down the broken leaf of a Hart's Tongue Fern (*Scolopendrium vulgare*) I tasted it."

Among the plants gathered were the Wood Sorrel (*Oxalis acetosella*) with its red-jointed creeping roots, trefoil leaves, and white delicately veined blossoms, the Wild Garlic (*Allium ursinum*), the Yellow Weasel-snout (*Galeobdolon luteum*), and Marsh Red-rattle (*Pediculus palustris*).

The banks of the Exe, according to collections seen by the author, afford about 360 flowering plants, among which may be mentioned the Flowering Rush (*Butomus umbellatus*), and Marsh Orchis (*O. latifolia*). Near its source the three Heaths (*Erica cinerea, E. tetralix,* and

Calluna vulgaris) are very abundant, and about seven Orchids, two Sundews, the Bog Pimpernel (*Anagallis tenella*), the pale-flowered Butterwort (*Pinguicula lusitanica*), and two species of Cotton Grass. In a Devonshire bog I once noticed this plant growing in such abundance that I was able, in two or three visits, to fill a bag with the downy heads. I collected altogether about a pound weight. About forty years ago an old gentleman in the north of Devon ploughed some acres of boggy land which he sowed with the seeds of this cotton grass, which in due time produced an abundant crop. The old gentleman published a pamphlet in which he proposed that England should thus grow her own cotton. He sent examples to spinners for trial, but they declared that the fibre would not card. As the Exe approaches the sea, its marshy banks are lined with the Sea Purslain (*Atriplex portulacoides*, Lin.), the Wild Celery (*Apium graveolens*), and the Marsh Samphire (*Salicornia herbacea*). The Exe is a salmon river, and abounds in eels, roach, dace, and about twenty-five other species of fish. These vary according to the part of the river examined.

The Ouse, a small river of Sussex, is conspicuous for the number of beautiful plants which may be gathered on its banks in the neighbourhood of Lewes. The fishing is inconspicuous. The tide comes up to Lewes, from where it enters the sea at Seaford. The chalky soil favours the growth of certain plants. The marshes of Lewes possess almost all the marsh plants of the South of England, in addition to those characteristic of chalky formations. Hence the number is greater than is usually found. Among the less common may be mentioned the Fringed Water-lily (*Villarsia nymphæoides*), the intermediate Bladder-wort (*Utricularia intermedia*), the Water Violet (*Hottonia palustris*), the Brook Weed or Water Pimpernel (*Samolus Valerandi*), the White and Yellow Water-lilies (*Nymphea alba* and *Nymphea lutea*). The number of plants found in the vicinity of the river is greater than that I ever found in one locality, not less than 300 species. The number of insects also is very great; the number of species obtained during two years' collecting not being less than 700, and these by one collector. The fresh-water shells of the Ouse and of the brooks that run into it are very numerous, such as *Limnea stagnalis*, *L. pereger*, *L. palustris*, *L. glaber*, and two other species; *Planorbis corneus; Carinatus, vortex, nitidus, albus,* and *nautilus; Bythinia tentaculata,* (and *Leachii*, lower down the river in a ditch); *Cyclas cornea* and *rivicola; Paludina vivipera; Pisidium,* two species; of *Succinea* two species, *Unio pictorum* and *ovalis; Anodonta cygnea;* in truth, nearly all the British land and fresh-water shells.

No better place for the naturalist to study botany, land and fresh-water shells, and entomology, can be found than Lewes. A little geology may be studied in the chalk quarries and features of the country.

The Medway rises near East Grinstead, in the south-east of Surrey, and flows through the centre of Kent, past the towns of Maidstone, Rochester, and Chatham. It is, perhaps, the most beautiful river of our south-eastern counties, and at one time greatly abounded in fish, which lurked in its deep black waters; but the numerous paper-mills, by discharging water contaminated with chloride of lime, have greatly destroyed them. In former times sturgeon was so exceedingly plentiful in the Medway, that the duty paid upon its sale to the Bishop of Rochester formed a considerable part of his revenue, although he received only a third share in that duty, as the King and the Archbishop were also entitled to a share. Hasted tells of a sturgeon taken near Maidstone, in 1774, which weighed 160 pounds.

Parts of the river are well stored with dace, carp, tench, chub, roach, and gudgeon, particularly above Maidstone. In the time of the Romans in Britain salmon was plentiful in the Medway. "Once on a time the Bishops of Rochester kept their pack of hounds by the riverside, and the monks gathered in the yearly vintage from carefully-tended grape vines in the vineyards of Medway valley; and bishops, monks, and laity fasted vigorously on a plentiful supply of salmon, bred and caught in Medway water; then there were no mills, no obstructions, or weirs to stop the ascent of the fish."

CHAPTER VIII.

OUR AQUATIC MOLLUSCA,—SPHÆRIUM, PISIDIUM, UNIO, PEARL MUSSEL, ANODONTA, DREISSENA, NERITINA, PALUDINA, VALVATA, PLANORBIS, LIMNEA.

THE fresh-water bivalves of Great Britain compared with marine shells and animals, those of our rivers, lakes, and ponds, are not showy in colour, and to many persons are much less attractive; yet they need only to be minutely examined and their habits studied to convince the intelligent observer, even if he be no naturalist, that, like all natural objects, they abound in beauty and interest, and richly reward those who have time and patience to devote to their study.

A great advantage attaches to this pursuit, as it would be difficult to find any locality in the country where it may not be carried on with more or less success. Every spot is at no great distance from some stream, canal, or pond, even if nothing worthy of the name of river or lake be within walking distance. The despised weedy horse-pond may contain some treasures for the beginner, new to him at least, if not to his neighbour or friend, the old collector of specimens. This applies forcibly to those young persons who are brought up in the country, and whose love for fishing in the ditch or brook is so well known to every parent or guardian. The young have a natural love of finding something which is new to them, which requires trouble to look for, and involves labour and time spent in the open air. True this natural inclination seldom gets beyond picking out of the water what may be for the moment attractive, only to cast it aside as useless or to break it up in ignorant thoughtlessness; but the teacher or parent may usefully direct the natural taste, and turn what was only an unthinking love of play into a habit of observation and a most pleasurable way of training the mind, with advantages of an almost priceless kind to the future life of a young man, though he be not destined to become a naturalist, or have that turn of mind which might make him in the long run one of the scientific men of his day. In no branch of knowledge can a habit of observation and analogy be more pleasantly, easily, and certainly taught than by encouraging and directing the young in their country rambles to look for, examine, and collect the shells, with their living inhabitants, pebbles, or insects of their neighbourhood,

These fresh-water bivalves are distinguished by the leaf-shape of the gills, which are in pairs on both sides of the body. They have no distinct head; the foot is tongue-shaped, and, in some instances, can be considerably elongated; it is used by the animal for creeping along or fastening itself to other objects "by a lyssus or bundle of muscular threads." Like the class *Monœcia* among plants, both sexes are comprised in the same individual; the respiratory organs are the gills. The mantle is divided into two lobes; this envelopes the body, and is shaped like the shell which covers it. The mantle often has tentacles or feelers at the edges; these are short filaments. The mouth, as there is properly speaking no head, is contained in the fold of the mantle. Some species have rudimentary eyes placed in the "interstices of those filaments where the mantle is open." The fresh-water bivalves of Great Britain comprise in all three families. The first is the *Sphæriidæ*, of which the body is sub-globular; in front the mantle is open, and the back or inner side forms a cylinder frequently divided near the opening into two *tubes*. If there are two tubes, one of which is longer than the other, it is used for respiration and nutrition, the shorter for excretion. The outer edges of the mantle are not furnished with filaments. The mouth is between the adductor muscle and the *sole* of the foot, and has two small lips. The foot is thin, of a wedge form, and can be considerably extended. The shell is thin and oval, and though the valves are of the same size they differ in shape, being unequally convex on the outer surface. The shell within is lined with nacre or mother-of-pearl, and outside has a skin or epidermis; this latter is generally removed by the collector as hiding the beauty of the colour underneath, but to the scientific collector it is a defect to have a specimen deprived of its natural skin. The hinge, which is an important part of the animal's shell, has upper and under teeth, which close and open at the pleasure of the inhabitant. A muscular cord or band fastens the valves together, generally external, but sometimes within the hinge. This family produce their young alive, but they retain them for some time between the mantle and gills. Sometimes they have a byssus or thread proceeding from the foot by which they can suspend themselves. They crawl along on their foot like a leech, and sometimes float in the water. They feed on animalculæ, but in winter remain torpid. The *Sphæriidæ* are like the marine *Kelliadæ*; but the *beaks* of the shells in the latter are more pointed, and the ligament of the hinge is internal. This family has been carefully studied by the Rev. Leonard Jenyns, and there is much information about it in the "Transactions of the Cambridge Philosophical Society," 1832. Among French conchologists M. Bourguignat has produced an elaborate essay on the recent and fossil species of *Sphærium* or *Cyclas* found in France, published in the *Mémoires de la Société des Sciences physiques et naturelles de Bordeaux*, tome i., 1854. One species only not found in Britain has been noticed by him, the *Cyclas solida* of Normandy. This, Jeffreys says, forms a link between *Sphærium* and *Cyrena*. M. Bourguignat entitles it generically *Cyrenastrum*. The *Cyrena* or *Corbicula fluminalis* is found frequently in the upper tertiary beds. The other genus of this family is *Pisidium*, minutely described by Dr. Baudon in his monograph published at Paris in 1857. Mr. Jeffreys gives the preference to the older name *Sphærium* rather than *Cyclas* to genus 1 of this family; the former name was given by Scopoli in accordance with the spherical shape of the animal. The body in this genus is nearly equilateral, and the mantle has a double tube. The shells differ slightly in concavity, and the beaks are near the middle of the back edge. The first species of this genus is *Sphærium corneum;* it has a sub-globular shell, nearly equilateral, it is glossy, of a horn colour, often with paler bands, which indicate periods of growth, sometimes with rays of faint brown from the beaks to the front edge, with indistinct longitudinal lines, under a magnifying glass appearing reticulated. The epidermis is thin, the beaks central, the ligament short, a strong hinge with a double cardinal tooth in each valve, two lateral teeth in the right and four in the left valve. The inside of the shell is bluish-white. The length is 0·35, the breadth 0·45. There are several varieties; one is of a straw colour, and is smaller and rounder. One, again, is more oval, and paler in colour. Another has the shell sub-triangular, with transverse striæ. It inhabits slow rivers, ponds, and ditches throughout the country. *S. rivicolum,* the second species, may be distinguished from *S. corneum* by its greater size, the length being 0·7, breadth 0·9, and oval shape; the ligament is very conspicuous, and the ridges on the shell are strong. It inhabits the same localities.

SPHÆRIUM RIVICOLUM.

S. ovale, the third species, has the shell oblong, compressed less equilateral than the preceding; it is thin and semi-transparent, yellowish, with sometimes a darker tint, and zones of growth, occasionally some faint rays, towards the margin fine concentric lines, the anterior side rounded, the posterior truncate sloping to the margin, which is curved and sharp, the beaks are small, the ligament long and narrow, the inside is ashy-white, the teeth resemble those of *corneum*, but are very small. The scars left by the adhesion of the animal to the shell are faint. Its habitat is the Exemouth, the Paddington Canal, and the ponds and canals of Lancashire. "A living specimen taken in February, and kept in a vessel by itself, gave birth to several young, which were immediately very active, and used their long foot like a leech. They seemed to be fond of nestling under their mother for the sake of shelter or shade." The shells are slightly iridescent.

S. lacustre, the fourth species, is 0·3 long, and 0·4 broad; the body is whitish, tinged with rose colour. The foot is nearly twice as long as the shell, which is nearly round; the mantle is fringed with gray, the epidermis is very thin, the shell is very thin, horn or grayish colour, with sometimes darker zones, and is iridescent. The anterior and posterior sides are cut off and slope towards the front margin. The inside is bluish-white. They generally inhabit lakes, ponds, and stagnant water in England, Wales, and Ireland. These animals are interesting in their habits in captivity; they possess the power of spinning transparent threads, by which they attach themselves to water plants, and have other interesting manners which space will not permit us to describe.

The genus *Pisidium*, the name signifying pea-shaped, was established to divide from *Sphærium*, the smaller species, which have only one tube or siphon, and whose shells are less equilateral. The general shape of the shells and size of the beaks are said by Jeffreys to be the only reliable method of

distinguishing species so nearly allied. He says:—
"Size, substance, sculpture, and lustre are not of much account, as they mainly depend on the chemical ingredients of the water inhabited by these molluscs as well as on their supply of food." He says that he has to the best of his ability worked out the subject of species, and submitted his labours to the criticism of naturalists. As an exemplification of his labours he states that his "cabinet contains no less than 274 parcels of *Pisidia* collected from various sources during thirty or forty years. Among the British and Continental specimens he has examined he can only recognise six distinct species. The British *Pisidia* are five in number, distinguishable mainly by their forms. 1. *P. amnicum*; 2. *P. fontinale*, triangular; 3. *P. pusillum*, oval; 4. *P. nitidum*, round; 5. *P. roseum*, oblong. The body of *amnicum* is grayish-white, rather transparent, tube short, sub-conical, obliquely truncate at its orifice, foot broad at its base, abruptly pointed, and very extensible, mantle bordered with grey. The shell solid and glossy, horn-coloured, length 0·3, breadth 0·375. The epidermis is thick, beaks prominent, ligament short, inside blueish-white and nacreous. It is the largest kind of *Pisidium*. The epidermis of *fontinale* is thin, beaks acute, inside white. In many respects similar to preceding species. There are several varieties of both. They inhabit slow streams, ponds, and canals in all parts of the kingdom.

The body of *pusillum* is whitish, with occasionally a tinge of yellow or red, the foot is longer than the shell, the mantle bordered with reddish-gray. The shell is less glossy, and it is striated irregularly, but finely, denoting the stages of growth, the colour is cinereous or yellowish-white, the length is 0·175, breadth 0·2; it inhabits mossy swamps, ditches, and drains, and is generally diffused; the variety *obtusalis* is less frequent.

The body of *nitidum* is "whitish, with sometimes a tinge of yellow, caused by the colour of the liver," the foot long, the mantle bordered with gray. The shell is thin and iridescent, striated concentrically; the colour yellowish-white, or light horn; the epidermis "a mere film;" the hinge, as in the two last-named species, is short and strong; its habitat is also similar. There is a fine variety, nearly half as large again, of a lemon colour.

P. roseum has the body of opaline colours, the foot long and semi-transparent; the shell is glossy, yellowish-white, and the epidermis is very thin; the inside of the shell is nacreous-white, the hinge-line is nearly straight, the cardinal teeth are very small, the lateral teeth are sharp at the edges, otherwise not well developed; the scars, when the animal is removed, are scarcely visible; the length is 0·1, the breadth is 0·15. It inhabits localities similar to the preceding species.

The *Unionidæ* are often called fresh-water mussels. Some species are ovoviviparous, others oviparous, like most molluscs, and are believed to be monœcious. They live in rivers and large pieces of water, and with their large fleshy foot they move to considerable distances. In summer, when the streams are shallow, they bury themselves in the mud; they do so also in winter. Their food consists of *entomostraca* and minute animals. They much resemble mussels, but the chief difference appears to be in the position of the head and ligament, which is external in the *Unionidæ* and internal in the *Mytilidæ*, and the beak is in the former further from the anterior end, while in the *Mytilidæ* it is nearly terminal. There is a wide field still open for discovery in the habits of the *Unionidæ*, as not much has been done in detail by British naturalists.

Unio tumidus, Philipsson, has the body grayish, the mantle bordered with white; the excretal orifice is a short tube, brownish, with sometimes purplish streaks. The branchial orifice is mottled with orange-brown, the foot milk-white, with a pale orange tint thick and broad, the gills pale gray. The length of the shell is 1·5, the breadth 3. The form is oval, convex above, wrinkled, glossy, yellowish-brown, the epidermis thick, the beaks slightly incurved at a distance of one-fourth from the anterior side, the umbonal region strongly plaited in a wave-like manner, the folds sometimes rising into tubercles, the ligament strong and short; the anterior side is rounded and sloping towards the front, the posterior side slopes to a wedge-like point, the outside is white and narrow, with a tinge of blue; the right valve on the anterior side has a broad, thick bifid tooth, slightly bent, grooved so as to make its crest notched, having on its posterior side a groove to receive the tooth or lower valve; the left valve has a similar tooth wedge, shaped to fit the double tooth of the right valve, into which it locks. This valve has also a plate on the posterior side, mortised into the groove above mentioned. The length is 1·5, the breadth 3.

There are two varieties of this species. The first has a green epidermis with yellow rays, the latter is dark olive-brown. This species inhabits rivers, canals, and ponds as far north as Yorkshire. It is one of our Upper Tertiary fossils. Much larger specimens occasionally occur. One is mentioned as having been taken in Leicestershire, $4\frac{1}{2}$ inches in breadth and 2 inches long, the weight being above 3 ounces. (*Zoologist*, 1857.)

A single *tumidus* has been known to lay 1,500 eggs in a few days; they are deposited in clusters of about 100. *U. pictorum* has the body red, with a grayish tint, the mantle bordered with brown, orifices like the last described, the foot reddish or yellow-white, tongue-shaped, the shell is long and compressed underneath the epidermis; in this and the last-named species the shell is cream-white. The beaks are a little incurved, and placed at a distance of between one-fourth and one-fifth from the anterior side; the umbonal region is not so prominent nor so strongly wrinkled as in the preceding; the teeth are finer, sharper, and more erect, the muscular scars distinct, the pallial scar faint, owing to the greater thickness of the nacreous lining; the length is 1·33, the breadth 3. The varieties are very numerous and diverse, having peculiarities to which significant names have been given. This species inhabits rivers, ponds, and canals.

Unio Margaritifera has the body dirty gray, with sometimes a tint of flesh colour, the mantle bordered below with brown, and above with white cirri, oblong and dark-brown foot, large, tongue-shaped, grayish-yellow or dirty red, gills grayish, labial palps broader than long, and united for two-thirds of their length. The shell is oblong, 2·4 long, and 5 broad. It thus differs from all the others in size. It is found in mountain rivers and streams throughout the British Isles. It is also darker in colour; the erosion of its umbonal region, and especially in the posterior teeth, being scarcely developed. The lining of mother-of-pearl is equal to half the thickness of the shell; the shell is dull white under the epidermis.

The rivers of Scotland produce finely-formed pearls in the mussels of *Unio Margaritifera*. A pearl which was found in the river Conway, in Wales, was presented by Sir Richard Wynn, of Gwyder, to Catherine of Braganza, Queen of Charles II. This pearl was placed in the king's crown, and it is said that a Scottish pearl, called the Ythan Pearl, is in the Scottish crown.

The famous Pennant, in 1769, made a tour of Scotland, and in speaking of the river Tay says:—"There has been in these parts a very great fishery of pearls got out of the fresh-water mussels from the year 1761 to 1764. Ten thousand pounds worth were sent to London, and sold from 10s. to £1. 16s. per ounce. I am told that a pearl had been taken there which weighed 33 grains. But this fishery is at present exhausted, from the avarice of the undertakers; it once extended as far as Loch Tay."

The Rev. Mr. Bannerman in his account of Cargill, a parish on the banks of the Tay, says:—"About twenty years ago there was a great demand for pearls, and many were occupied in fishing for them." This was about a hundred years ago. Afterwards the demand for pearls became less, and the fishing fell off.

Miss Drummond, a lady of Perthshire, has or had a necklace, the pearls in which were taken from the Tay, and are of considerable value. This had been in the family for several generations. "The size and shape of these pearls are not to be equalled by anything of the kind in Britain."

The Rev. Mr. Robertson in his account of the parish of Callander, in western Perthshire, says:— "In the Teath are found considerable quantities of mussels, which some years ago afforded a great profit to those who fished them by the pearls they contained, which fetched high prices. Some of the country-people made a hundred pounds a year by that employment. This fishery was soon exhausted, because only the old shells, which are crooked, in the shape of a new moon, produce pearls of any value."

Pearls have been found in the bed of the river Devon, which rises in the Ochil hills. The Esk, in Forfarshire, at one time had on its banks a fishery for pearls which was productive. Some of these were so valuable as to produce £4 at the first market. The Rev. Mr. Jamieson, of Forfar, in his account of the parish of Tannadice, says "one was found nearly as large as the ball of a pocket pistol." This fishery also was abandoned.

The river Cluny, in Aberdeenshire, is stated by

Mr. Richie to produce pearls, and some years before he wrote, a Jew is said to have employed people to fish for them, and a good many were taken. In the Ythan, a stream of Aberdeenshire, many pearls were formerly found by the shepherds who fed their flocks on its banks. In dry summers numbers of small pearls were said to be found in the mussels in some of the streams in Kirkcudbright, in the south of Scotland. This fishing for pearls in Scotland was discontinued for many years. But again Scottish pearls became fashionable, and the search for them revived.

A writer in *Land and Water* about ten years ago wrote:—" During the season a pretty large number of persons engage in the pearl fishery, which is open to all. The mollusc is obtainable at none of the serious risks attending the pearl fishery of the East; for, instead of lying at a great depth in the bottom of the sea, it is usually found among the mud in the beds of rivers, which the summer droughts render shallow and clear, so that it can easily be perceived by the eye and almost reached by the hand. In this way boys and girls can perform the work of pearl fishing as easily as their elders. The 'herd laddie' on the hill-side can go down to the neighbouring stream, and if the water be limpid and shallow he can pick up the mussel without the aid of any apparatus, or if it is deep he has but to reach down a stick and insert the point of it betwixt the open shells, which instantly closing like a vice upon the intruding body, the mussel can be thus lifted up."

The same writer says he knows of many fine pearls

being taken in the vicinity of Perth, which brought high prices. The Earn, a tributary of the Tay, is also amous for pearls, and some families living on its banks derive their living chiefly from this source. A dealer in Edinburgh some years ago is said to have asked £350 for a necklace of Scottish pearls, the single value of each pearl being estimated at from £5 to £90. The best Scottish pearls have a delicate pink hue. The mussel is found singly in the muddy bottoms of rivers, and not in beds, like oysters, and the pearls are only bred in old shells.

Spruel, an old writer, states that a " birthy shell " should be wrinkled and nicked like the horn of a cow, "for the more nicks or wrinkles in the shell, the older and better the pearl is. Smooth shells are barren."

Some of the rivers of Ireland were also famous formerly for pearls. Speaking of the value of Irish pearls two hundred years ago, Sir Robert Redding, F.R.S., wrote to Dr. Leslie, 13th October, 1688, as follows :—" Being in the north in August last, and calling to remembrance your desire to have some of the mussel shells sent you wherein the pearls were found, I have sent four or five of the shells, and a few pearls taken out of the river near Omagh, in the county of Tyrone. The poor people, in the warm months before the harvest is ripe, whilst the rivers are low and clear, go into the water—some with their toes, some with wooden tongs, and some by putting a sharp stick into the opening of the shells, take them up. And although not above one shell in a hundred may have a pearl, and of those not above one in a hundred be tolerably clean, yet a vast number of fair merchantable pearls, and too good for the apothecary, are offered to sale by those people every summer assize. Some gentlemen of the country make good advantage thereof; and myself, while there, saw one pearl, bought for fifty shillings, that weighed thirty-six carats, and was valued at forty pounds; and had it been as clean as some others produced therewith, would certainly have been very valuable. A miller took out a pearl which he sold for four pounds ten shillings to a man that sold it for ten pounds, who sold it to the late Lady Glenarty for thirty pounds, with whom I saw it in a necklace. She refused eighty pounds for it from the Duchess of Ormond.

"That part where the pearl lieth is in the toe or lesser end at the extremity of the gut, and out of the body of the fish, between the two films or skins that line the shell.

"I believe that this pearl answereth to the stone in other animals, and certainly like that increaseth by several crusts growing over one another, which appeareth by pinching the pearl in a vice, and the upper coat will crack and leap away: and this stone is cast off by the mussel and voided as it is able, and many shells that have had pearls in them are found now to have none; by these instances the shells that have the best pearls are wrinkled, twisted, or bunched, and not smooth and equal as those that have none, as you may observe by one of the shells herewith sent, of a lighter colour than the rest: this shell yielded a pearl sold for twelve pounds. And the crafty fellows will guess so well by the shells that, though you watch them never so carefully, they will open such shells under the water, and put the pearls in their mouths or otherwise conceal them. That same person told me that when they have been taking up shells, and believed by such signs as I have mentioned that they were sure of good purchase, and refused good sums for their shares, that yet they have found no pearls at all in them.

"Besides the river near Omagh, other rivers which empty into Lough Foyle produce pearls; so also the Suir, running past Waterford, the Lake of Killarney, in Kerry, and others."

In Connemara they are still taken, and sold to tourists for a trifle. This is chiefly at Oughterard and Letterbrack. The pearls at the former place being chiefly got in the Owenfough, a river that runs through that village; and the latter chiefly at Dawros, the river that runs westward into the Atlantic from Kylemore Lough.

The Irish pearls, like the Scotch, are said to have a slight pinkish tinge. The Dublin jewellers do not value them, as they say they will not cut. Some are very large, one as much so as a No. 5 Colt pistol bullet; but it is not of a good colour, being somewhat brownish. The waste in this search is very great, for hundreds are opened which have no pearls, and many young ones uselessly so, as they could never have had any.

The largest of British fresh-water shells is undoubtedly *Anodonta*.

There is considerable difficulty in distinguishing

between some species of *Unio*, and still more in the case of *Anodonta*.

A. cygnea has the body gray, with yellowish or reddish tint, the mantle bordered with brown, the foot broad, dirty yellow lined with orange or red, the gills gray with a gauzy texture. The shell is oblong, yellowish-green, thin, moderately glossy, transversely and irregularly grooved by the lines of growth, wrinkled in the same direction on the posterior and lower sides; epidermis thin, beaks straight, ambonal region compressed, strongly plaited; ligament rather long, strong, and partly concealed, the inside pearly-white and iridescent.

Of this species there are several varieties, all very distinctly marked, and by older authors called species. The length is about 2·75, the breadth about 5·35. It is an Upper Tertiary species. The young have tri-angular pearly shells, easily confused with *Cypris* or smaller *entomostraca*.

A. raticeta is 2·1 long, and 3·5 broad. The shell is oval, compressed, but not so thin as the last-named species; the colour is olive-green or brown. The epidermis is thicker than in *A. cygnea*.

The family of *Dreissenidæ* contains but one genus. The shell is equivalve, oblong, triangular, inequilateral, ventricose, and covered with a horny epidermis, the beaks are at the anterior end, the ligament is internal, and the inside porcelain-white; the hinge has sometimes small cardinal teeth. Below the beak in each valve is a hollow plate for the reception of the anterior muscle.

The genus *Dreissena*, which may be called the

fresh-water mussel, is gregarious, and they attach themselves by a strong byssus to extraneous substances. They can live a long time out of the water, and are believed to have come from Tartary.

D. polymorpha (Palla) is 1·4 long, and 0·6 broad. It inhabits slow rivers, canals, and lakes. It has been even found in the water-pipes in London. The shell is oblong, rising in a sharp keel in the middle of each valve and flattened below, pointed at the end or beak, and gradually but obliquely widening towards the front, solid but not glossy. Beneath the epidermis it is purplish-brown. The beaks are terminal.

We shall now speak of the *Univalves*.

The body is conical, the mantle forms a single lobe, the head is usually distinct and furnished with tentacles, of which the upper pair has two eyes, either at the tips or base, on separate stalks, the foot is a muscular disc by which the animal crawls or floats. Some species are hermaphrodite. The respiratory system consists of gills or lung-like organs, the former being possessed by aquatic kinds. The shell is conical or spiral, and covers the whole or most of the body.

The order *Pectinibranchiata* has the body of spiral form, and has a single gill within the mantle, the shell is external. Three families of this order inhabit the fresh waters of this country—1. *Neritidæ*, 2. *Paludinidæ*, 3. *Valvatidæ*. All these fresh-water snails have two tentacles and eyes at their base. The shells have an operculum and an epidermis.

Of the family *Neritidæ*, there is one species of the genus *Neritina* (Lam.); it is the *Neritina fluviatilis*

(Lam.). The shell is triangular-oblong, the operculum calcareous and solid. This genus inhabits waters which have a stony or gravelly bottom. They are often encrusted with a calcareous matter; they are sluggish in their habits, but their tentacles appear to be always in motion. They feed on vegetables. Their eggs are generally deposited on the shell, and carried there till hatched. The eggs are round, of a yellow colour, and provided with a leathery covering which splits in two when the fry are excluded, the upper half being detached, and the other part left adhering to the parent-shell. Moquin Tandon says the eggs are deposited in a cluster of fifty or sixty. The length is 0·35, the breadth 0·25. They are oviparous.

The family *Paludinidæ* are more active in their habits than the last-named division; they are ovoviviparous. They occasionally float. They are the largest of our fresh-water univalves. The males are smaller than the females.

Paludina (Lam.) has the eyes on short pedicles; the operculum is horny. *P. contecta* has the body dark gray or brown with yellow specks, the head small, the eyes round and black, the foot cloven. The shell is conical, yellowish, with sometimes a green or brown tinge with brown bands, a thick epidermis, a thin operculum. It is marked with concentric striæ and lines of growth. It adheres to stones or wood, but when touched falls off. It is sometimes 2 inches long, and $1\frac{3}{4}$ inches broad. *P. vivipara* has the body of a darker colour, the snout broad, and the tentacle blue-black with yellow spots. The male is

P

smaller than the female. The shell is oval, not so glossy as *P. contecta*. The variety *unicolor* is without bands. This species and the preceding inhabit slow rivers and canals.

Bythinia (Gray) has the eyes sessile, the operculum testaceous and solid, and the nucleus nearly in the middle. The chief difference between this and *Paludina* is that *Bythinia* is oviparous. The eyes are sessile instead of being placed on stalks. Although the name would indicate that they inhabit deep water, yet they generally frequent shallow streams, canals, or ditches. *B. tentaculata* has the body nearly black on the upper part, while below it is yellowish; the head is small, the eyes large. The form of the shell is subconical, or oval, of an amber or brownish colour, opaque, and glossy; the length is 0-5; the breadth, 0-25. The varieties are *ventricosa*, with a white shell; *decollata*, with the upper whorls wanting; *excavata*, with the whorls rounded, and suture deeper. It is said to feed partly on animal substances. *B. Leachii* (Sheppard) has a conical shell, thin, and semi-transparent, horn, or amber colour, very thin epidermis; the operculum is flat, and nearly round; the length is 0·25; the breadth, 0·2; the body is grayish-white, with yellow specks.

Hydrobia (Hartmann) has the eyes on tubercles, and a thin, horny operculum. These small molluscs live chiefly in fresh water, but *H. ventrosa* is found in estuaries, and pools near the shore. In Britain there are only two species of Hydrobia, *similis* and *ventrosa*. In both the shell is semi-transparent, usually horn colour, but in *similis* sometimes white, as also in the variety *pellucida ; similis* is 0·15 long, and 0·1 broad. . *Ventrosa* is 0·2 long, and 0·125 broad. This little shell is abundant, and has been called by several names. It is said to be the *Cyclostoma acutum* of Draparnaud. It chiefly differs from *H. similis* in having the spire long, the suture not channelled, and on the umbilical clinch being much smaller. The varieties of *H. ventrosa* are *minor*, of which the spire is small; *decollata*, of which the shell is eroded, and the spire truncate; *ovata*, having only four whorls, and a short spire; *elongata*, having a long spire, and sometimes 8 whorls; *pellucida* has a white, transparent shell. It is found throughout England and Wales, and even in Ireland.

The genus *Valvata* are very minute animals. The body of *V. piscinalis* is clear yellowish-gray, with small white spots; the shell is of a depressed conical form, opaque, brownish-yellow, with 6 whorls, the last being nearly half the shell. The varieties are *depressa*, with shell more depressed; *subcylindrica*, with the spire more produced, and flattened at the top; *acuminata*, with a sharp-pointed spire. The usual length of *piscinalis* is 0·25 ; the breadth 0·275. In *V. cristata* the body is dark, with black spots. The shell forms a flat coil, concave, rather solid beneath, semi-transparent yellowish, or gray horn-colour, very thin epidermis, 5 whorls, the last greater than all the others. The operculum is formed like an inverted potlid. These species inhabit lakes, canals, and ditches throughout Britain.

Those species of the order *Pulmonobranchiata*, or with a lung-like gill, which have no operculum have the sexes united in one, but "require fertilization by another individual." Those species which are aquatic require atmospheric air, but in Britain three-fourths of this order are terrestrial. Most of them are herbivorous. The principal freshwater genera are *Limnæa* and *Planorbis;* in the latter the body is long, and, in correspondence with the shell, twisted in a flat coil. The tentacles are very long, the foot short, attached to the body by a stalk. The shell corresponds in form to the body. The animal emits when irritated a red fluid. It has the vital organs on the left side, and the spire of its body is coiled in the opposite direction, *i.e.*, from left to right. The form of the shell resembles an Ammonite. The body is much smaller than the shell. Some of the smaller species of *Planorbis* shut themselves up in their shells during the dry weather until the return of rain. They all frequent stagnant or slow-running water. The eggs are in a globular bag, fixed to a stone or stalk of a plant. As they are sluggish the shells are apt to become encrusted with mineral or vegetable deposit. *P. lineatus* (Walker) has the body reddish-brown or purple, with black speck eyes, small foot broad. The shell is quoit-shaped, the upper more convex than the lower side. It has 4 whorls; inside the last whorl are rows of curved plates, arranged on each side across the spire. The length is 0·65; the breadth 0·2. *P. nitidus* (Müller) is shaped like the last, but flatter, and with more of the spire visible. The shell is thin, glossy, and prismatic, and of a light yellowish-horn, or gray, and sometimes reddish colour. The epidermis is very delicate ; the whorls are 4 or 5 : the last covers half of the preceding. It does not lay more than six eggs. The shell is often infested with egg cases of a water insect, or coated with confervæ.

P. nautilus (Linné) has the shell quoit-shaped; the epidermis is rather thick. There are 3 whorls, depressed above, the last exceeding in size the rest of the shell; the length is 0·035; the breadth, 0·1. The variety *cristatus* has the transverse ridges stronger, and the periphery notched by them. It is found on aquatic plants in marshes from Zetland to the Channel Isles. It lays from three to six eggs. Its small size and dull appearance, with its large umbilicus, distinguish it from the foregoing species. *P. albus* has a dirty gray-coloured body, with black specks; the eyes are small. The shell is flat, depressed in the centre; the epidermis is thick, and sometimes bristly. *P. glaber* (Jeffreys) has a yellowish-gray body, rather short tentacles, cylindrical, and ending in a blunt point; the foot is rather broad; the size is 0·05 long, and 0·15 broad. It is also found on aquatic plants, but it is not generally distributed. It is an upper tertiary fossil. It is smaller than *P. albus*. *P. spirorbis* (Müller) has the body purplish-gray or reddish-brown, with black specks on the foot. The shell is glossy, brownish horn colour, and the epidermis is thin; there are 5 or 6 whorls, gradually increasing in size; the length is 0·04; the breadth is 0·25. The variety *carinatus* has a smaller shell, with one whorl less than usual. It is found on plants and grass from the Moray Frith district to the Channel Isles. Like some other species, in dry weather it retires within its shell, and closes it with a yellowish lid until the rain comes.

P. vortex (Linné) has a thin compressed shell, with 6 or 8 whorls, gradually enlarging; the length is 0·05; the breadth 0·3.

The variety *compressus* is still flatter, and the shell thinner. It lays from ten to twelve eggs. *P. carinatus* (Müller) has the body a deep reddish-brown, with a yellow tint underneath, marked with fine black spots. The shell is compressed, concave above, and slightly convex below; the epidermis is thin; the length is 0·1, and the breadth 0·5. The variety *disciformis* is flatter and thinner, of a yellowish colour; the last whorl is larger in proportion to the others. It inhabits marshy and stagnant pools in the home and eastern counties, as well as in several northern and southern counties. *P. complanatus* (Linné) has the shell concave above, and slightly convex below, solid and opaque, yellowish horn colour, with a tinge of brown; length, 0·125, breadth, 0·6. The variety *rhomboides*, the shell smaller, rather more convex above, and deeply umbellicated below. Variety *albidus*, shell whitish, or colourless. It inhabits marshes, ponds, and standing waters throughout the Kingdom, highly polished; the epidermis is very thin. The whorls are 6 or 7 in number, convex, but slightly compressed. It is found from the Moray Frith to the Channel Isles, but is local. It is an Upper Tertiary fossil. *P. corneus* (Linné) has the body dark-red, or nearly black, greyish beneath, with black and gray specks on the upper part; the tentacles are long and curved with blunt tips; the eyes a moderate size. The shell is rather concave above and nearly flat below, of a whitish horn colour, with a reddish-brown tinge, irregularly striate by curved lines of growth marked with fine and close-set spiral striæ, more perceptible in the first whorls; the upper surface is pitted or impressed irregularly like cut-glass; the length is 0·35, the breadth 1. The variety *albinus* has a white shell. It is local, and occurs in many parts of England and Ireland. It is larger than any other European species of *planorbis*. When irritated it emits a purple fluid from its neck. On a warm day it floats on the surface of the water. It lays two or three capsules, each containing from twenty to forty eggs, which are excluded at the end of fifteen or

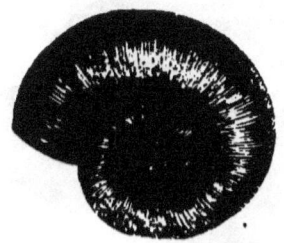

PLANORBIS CORNEUS.

sixteen days. The epidermis of the young shell is covered with fine down; its surface resembles velvet pile. *P. contortus* has a black body, tinged with red. The tentacles are slender, and the eyes small; the foot is broad, narrowing to a blunt tail. The shell is flat, with a deep depression or concavity in the middle, very concave below. The epidermis is thick; the length is 0·075; the breadth, 0·175. The variety *albinus* is nearly white. Its habitat is on water-plants in fresh ponds and ditches throughout Britain, as far north as Scotland, but it is local.

The *Physa hypnorum* has a lustrous dark gray,

brown or almost black, with sometimes a faint tint of blue covered with black or dark gray spots. The length is 0·5, the breadth is 0·2. It has 6 or 7 whorls. The shell is oblong, spindle-shaped, yellowish or reddish-brown, faintly striate by the lines of growth; has a dull greenish-coloured body, with small black and white spots. The foot is bordered with yellow; the tentacles are broad and flat, and the eyes small. The shell is semi-transparent horn colour, 4 or 5 swollen whorls, a small sharp-pointed spire. The length is 1·125; the breadth, 0·825. The other species

LIMNÆA STAGNALIS.

of the genus *Physa*, *P. fontanalis*, has an oval, thin, glossy, transparent shell, 4 or 5 swollen whorls, the last occupying more than three-fourths of the shell. The body is lustrous, dark gray, with sometimes a tint of yellow or violet. The varieties are *inflata*, half as large again as the usual size, the whorls angular towards the suture, the middle one more prominent; *curta*, with a very short spire; *oblonga*, with a long spire; and *albina*, with a milk-white shell. It is found on water-cresses and other aquatic plants in brooks, and also in still waters.

The genus *Limnæa* (Bruguière) frequents still waters. Some of the species appear transitional with *Physa;* some have the shell enveloped by an outer fold or lobe of the mantle. *Limnæa glutinosa* (Müller) has a dark gray body with a greenish-yellow tinge and light spots; the tentacles short, triangular with blunt tips. The shell is globosely oval, very thin, transparent horn colour, and highly polished. The variety *mucronata* has the spire longer, and the shell less globular. *L. involuta* (Thompson) is oval, yellowish-brown, a thin epidermis, 3 or 4 whorls. It inhabits a mountain stream near Killarney.

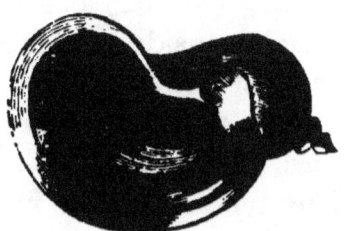

LIMNÆA AURICULARIA.

L. peregra has a yellow gray body with an olive tinge covered with black and light spots. The shell is obliquely ovate, thin, and rather glossy and transparent. It is 0·75 long and 0·425 broad. The variety *Burnetti* has the body broader than in the typical form; it is dark-olive, and the mantle nearly black with a few light spots. *Lutea* has a short spire and 3 or 4 whorls. *Lacustris* is smaller than *Burnetti*, but similar. There are several other varieties. *L. auricularia* is a well-marked species (Linné). *L. stagnalis* (Linné) has the body fawn or yellowish gray, with a reddish tint covered with small brown and white specks, long tentacles; the foot has a narrow edge of yellow. The shell is 2 long and 1 broad; there are 7 or 8 whorls convex, and bulging out in the middle; the colour like the last-named species; the epidermis is thin. There are several varieties, *fragilis, albinus*, &c. It is found throughout the Kingdom. *L. palustris* (Müller) has a dark-gray body, with a tinge of violet-brown, covered with black and yellow spots. The shell is oblong, of a dull colour, sometimes tinged with violet, sculptured as is *stagnalis*, but the spiral ridges are more prominent. There are several varieties. *L. truncata* (Müller) has a dark-coloured body, lighter on the lower side, with fine black specks; the tentacles are short and slender. The shell is oblong-conic, turreted, solid for its size, sculptured as in the last species. There are several varieties of this species. The typical size is in length 0·4, in breadth 0·2. *L. glabra* (Müller) has a dull gray body covered with minute spots; the eyes are placed on prominent tubercles. The shell is cylindrical, with 7 or 8 whorls, rounded, not very convex; the spire ends in rather a blunt point. It is sparingly distributed, and inhabits shallow pools.

The genus *Ancylus* of Geoffrey, has the body oval, conical, slightly twisted; the head is large. The respiratory pouch forms a short tube. The shell is hood-shaped. This genus is related to *Limnæa*. They inhabit rapid and still waters. The food of the *Ancyli* is algæ or confervæ, and decayed vegetable matter. They are also said to swallow a portion of sand to assist their slow digestion; they can live a long time without nourishment. *Ancylus fluviatilis* (Müller) has a dark slate-coloured body, with black specks; tentacles triangular at the base, eyes distinct, but not prominent. The shell is semi-oval, incurved like a helmet, the epidermis is thin. The shell is not glossy, the spire forms the beak, and is equal to about half a whorl. The length is 0·3; the breadth, 0·233. There are varieties, some larger, some smaller. It is found on stones from Aberdeenshire to the Channel Isles. *A. lacustris* (Linné) has a yellowish-gray body, with a greenish tinge. The epidermis is thick, and the beak sharp and ridge-like. The length is 0·25, the breadth, 0·1. Another variety is compressed, and is much larger and flatter than the type; one is milk-white, with a gray epidermis. This *Ancylus* is found on the under leaves of water-lilies, and other aquatic plants, as far north as Aberdeenshire. It has been taken in Staffordshire, and in the Grand Canal, Dublin. It is easily distinguished from *A. fluviatilis* by its habitat and the oblong shape of its shell. *Succinea amphibia, Pfeifferi* and *oblonga*, are little marsh-living animals whose shells are of a beautiful transparent amber colour; of nearly similar habits is *Vitrina pellucida*.

CHAPTER IX.

SIX MONTHS OF BROOK AND POND LIFE IN SUSSEX.

THERE is not a month of the year which does not yield to the observer of nature ample material. Even in the month of January, when frost so generally binds the soil, the pulses of vegetable life beat languidly, the insects that are in a mature state shrink into the smallest and warmest corners, and the dormouse and the lizard sleep, yet if the naturalist has vigour to walk abroad he will learn not merely from what he does find, but from what he finds not. With these thoughts in my mind on a January morning, I visited the brooks in the southern district of Lewes; and collecting with a stick some green mossy conferva from the surface, I found on examining it under the microscope and comparing it with engravings to be the spiral conferva, *Spyrogyra quinina*. A series of fine hairs, resembling spun glass, were divided into cylindrical cells of an emerald green colour. Attached to these confervæ were the bellflower animalcules *Vorticellæ;* beautiful transparent bells most lively in motion. The *Rotifer vulgaris* accompanied them; this is a much more

WHIP TAIL.
(A Rotifer.)

highly-organized being, having eyes, but these minute forms require an object-glass of about sixty diameters; a large number of the *Desmidaceæ* were also here.

February is not much more productive than January, but still our brooks afford the water-fleas, *Cyclops* and *Daphnia pulex*, and the *Hydra vulgaris*, a radiated

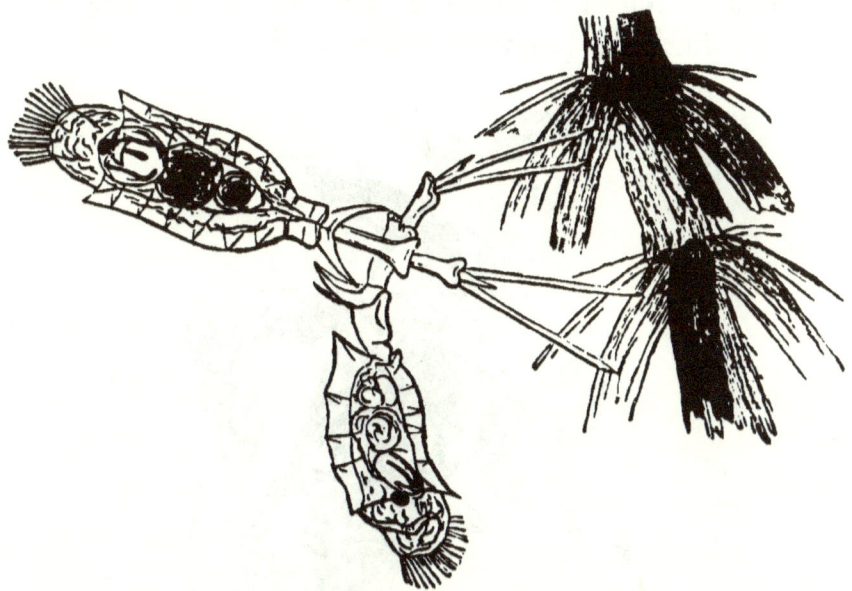

SKELETON WHEELBEARER.
(A Rotifer.)

animal, which attaches itself to minute plants, the empty shells of molluscs, and the empty cases of caddis-worms. The latter attach to themselves numerous shells, sticks, and other substances, which cause them to assume a most extraordinary appearance. In the summer they are found creeping, as in the cut, on aquatic plants.

The month of March is much more favourable.

The time of the singing of birds has come, and insect and plant-life begin rapidly to revive from the depressing season of winter. The frogs spawn and the tadpoles begin to develop.

There is much to interest in the metamorphosis of the frog if the spawn be kept in an aquarium and watched. On the banks of our brooks we may in March find the lungwort flowering, *Pulmonaria offici-*

CADDIS-WORM IN ITS CASE.

nalis; it is most common in Hampshire and Middlesex. The trout and numerous little fish begin to be lively in the water. The Mare's Tail (*Hippuris vulgaris*) is now conspicuous along the banks of our marshy cornfields. The Marsh Marigold (*Caltha palustris*), conspicuous by its golden yellow flowers, is abundant

LUNGWORT.
(*Pulmonaria Officinalis.*)

MARE'S TAIL.
(*Hippuris Vulgaris*)

Amongst fresh-water molluscs the various species of Limnæa, conspicuously *L. stagnalis, peregra, palustris,*

Physa fontanalis and *P. hypnorum*, the two latter being curiously reversed, very transparent little univalve

PETASITES VULGARIS.

shells, having a smoky black colour when containing the animal. In the Lewes brooks I found eight species of Planorbis, of which the largest and most conspicuous is *P. corneus*.

On the banks overhanging the river Ouse I found

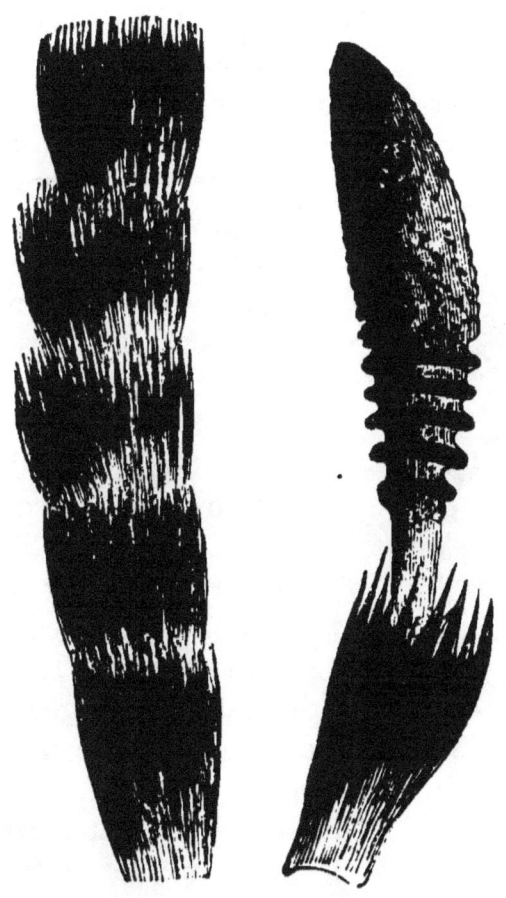

GREAT EQUISETUM.

the Butter Burr (*Petasites vulgaris*) in flower, and the Great Equisetum (*E. telmateia*); the leaves of the former do not appear till summer. The most luxuriant growth of the barren stems of the Equisetum comes also in summer. On the banks of the Ouse I once saw a wild duck's nest in this month.

April is still more conspicuous for the variety of its fauna and flora. The water-net enables us to capture a multitude of insects; among the most abundant are the water-scorpion (*Nepa cinerea*). This insect is of a gray-black colour, red underneath. Its front legs have great prehensile power, and with them it seizes its insect prey, which it sucks with its strong rostrum or beak. The rostrum contains three joints and four pointed bristles; two have a sharp blade, and have teeth towards their base. Of the others, one is like a thin needle, the other is provided with hairs directed backwards and forwards. The apparatus may be compared with that of a surgeon's cupping instrument, and its bite is painful to man.

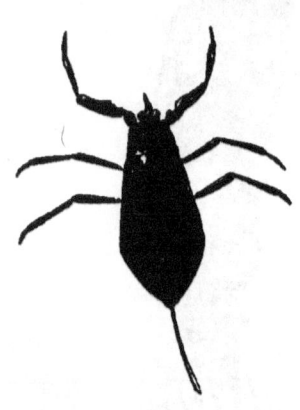

WATER-SCORPION.

One of the most remarkable of the inhabitants of these ditches is undoubtedly the water diving-spider (*Argyoneta aquatica*), which weaves a curious bell-shaped net beneath the water, to which it carries its prey. It fills this bell with air, which causes it to assume a silvery appearance. The air is carried down entangled in the hairs of the spider, suggests the method by which diving-bells among men were formerly supplied with air.

One of the most active and interesting of the insects found in the Lewes brooks is the *Notonecta*, which is very common. It has an oblong, narrow body,

WATER-SPIDER.

NOTONECTA.

convex above and flat below. Its head and eyes are conspicuously large. Its hind legs are very long and immensely powerful when used as oars, and enable it

THE HAUNT OF THE DRAGON-FLY.

to dart through the water at an amazingly rapid rate. At night it comes out of the water and flies well from one marsh to another. It has a most powerful rostrum, with which it can inflict a painful bite on man. This bite is fatal to insects from some supposed poisonous injection. The anatomy of these water-bugs is well worthy of close investigation. If the weather is fine, dragon-flies make their first

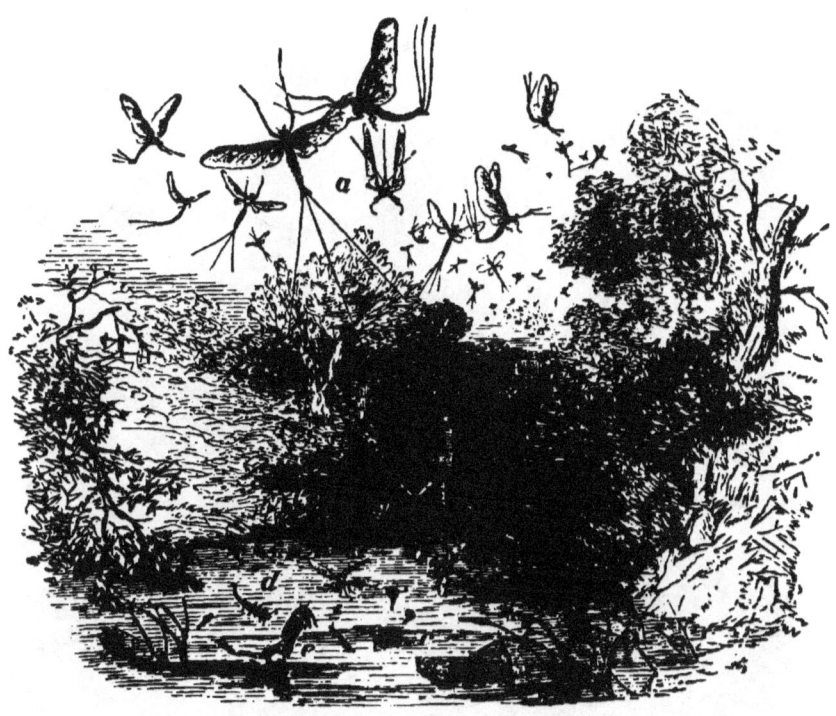

EPHEMERIDÆ OR DAY-FLIES.

appearance this month; they have been developed in a marvellous manner in the ponds and ditches, and excite our wonder and delight as they hover over the delicately-shaded pool to prey on the minute gnats and ephemeridæ.

If April was more productive than March, May is more productive than April in fauna and flora. The characteristic flies known as the *Ephemeridæ*, proverbial for their shortness of life, hover over the ponds and afford food to the swallows, trout, and larger insects. The Whirligig Beetles, *Gyrinidæ*, are exceedingly

WHIRLIGIG BEETLES AT SPORT.

abundant, and amusing by their lively motions. They swim with immense rapidity, and seem to have four eyes, two of which look down into the water, and two of which look into the air, so that the creature may be on its guard against its numerous enemies. The larva is a curious skeleton-like creature. They, in common with Dydiscus and Dragon-fly larva, are great favourites with the water-ouzel, as well as the duck tribe and water-rail. But the water-ouzel is not often found in Sussex. During the month of May, 1854, I was surprised by seeing the bearded tit playing among the reeds in a marsh near Lewes. Its favourite haunt and breeding-place is the fenny

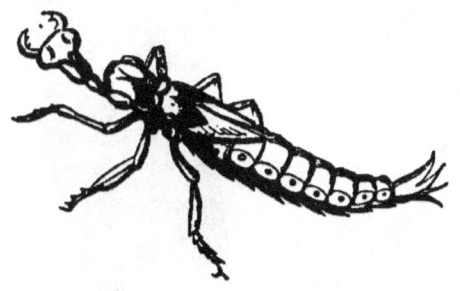

LARVA OF DRAGON-FLY.

district of Norfolk and Suffolk; but owing to the drainage of the fens it has been nearly extirpated. These are beautiful cage-birds, but they have no attractive song. The beautiful water-violet, *Hottonia palustris*, with its bunch of lilac flowers, was found rather abundantly in the Lewes brooks. This is a beautiful aquarium plant, and one that gives out a great deal of oxygen. Along the banks of the Ouse at the end of May I got plenty of the Yellow Iris, or Corn-flag, *Iris pseudo acoris*, which I remember from my early infancy as a familiar Devonshire plant, with its autumnal yield of pods of scarlet orange seeds,

YELLOW IRIS.

large as peas, which in some districts have been dried, roasted, and ground as a substitute for coffee. I do not know whether they possess any aromatic flavour or stimulating quality analogous to the theine of coffee.

The Water Plantain, *Alisma Plantago*, with its beautiful sprays of pink flowers; the Arrow-head, *Sagittaria sagittifolia*, an allied plant, distinguished by its leaves of an arrow-head form, and the flowering rush, *Butomus umbellatus*, which has somewhat the aspect of an umbelliferous plant, in having its flowers on separate stalks, coming out at the head of a main stem. The flowering rush is a gem amongst British water-plants. It is distinguished by its narrow grass-like leaves, and by its circular stem, which carries a large bunch of flowers at the top, beautiful by their bright pink colour and large size. The pistils are six in number and very conspicuous. It is subject to considerable variety in the colouring of the flowers, which range from the characteristic bright pink to white, and even purple. The plant is in bloom from June to September. It is widely distributed, being found alike in Italy, England, and Lapland. Water is everywhere necessary to its existence, and its mode of growth is recognised in the text, "Can the rush grow up without mire? Can the flag grow without water?" Job viii. 11. The first specimen I ever gathered of this plant was at Lewes, and the pleasure it gave me will never be forgotten. I had for seven previous years looked in vain for it.

Often in my study of the Linnæan system of botany I had wished for an example of the class *Eneandria*, containing nine stamens, but of which I was never able to find an example, except the common laurel. The flowering rush I found sometimes contained an even number of stamens, so that this was an unsatisfactory diagnosis. The Water-mint, *Mentha aquatica*, was very luxuriant in May, as well as the Brooklime, *Veronica beccabunga*, and the Water-speedwell, *Veronica anagallis*, a plant most valuable as an antiscorbutic.

WHITE WATER-LILY.

A rather rare plant at Lewes, but one which I found in one particular ditch, was the Buck-bean, *Menyanthes trifoliata*, distinguishable by its spike of pink flowers, most beautifully marked, and its trefoil-like leaves render it one of the most characteristic of our British plants. In Lancashire it is used as a bitter medicine, as a substitute for gentian, and in Sweden and Norway as a substitute for hops. The Water Ranunculus, *Ranunculus aquaticus*, that beautiful little floating plant, distinguished by its white flowers, and yellow stamens, and fine, thread-like,

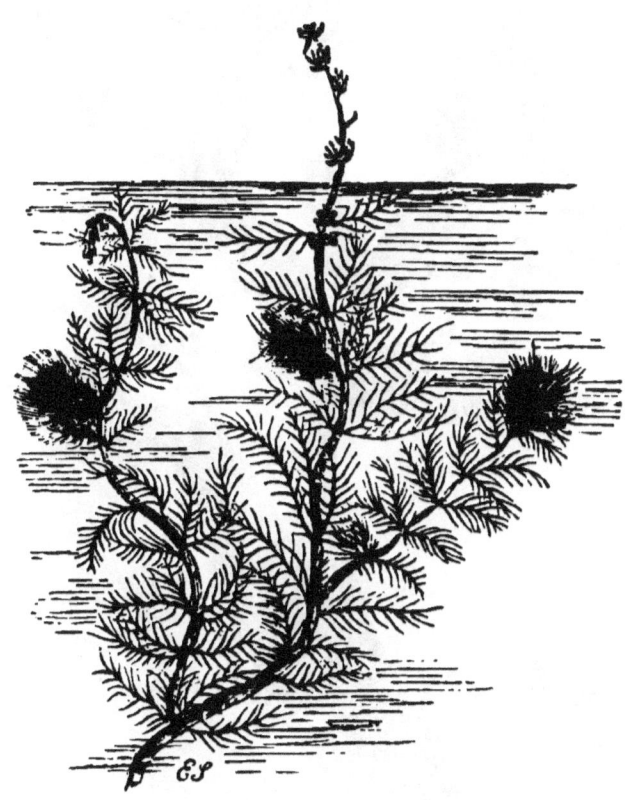

MYRIOPHYLLUM SPICATUM.

mossy leaves, form the favourite harbour for numerous water-snails and other mollusca, and innumerable insects, water-spiders, sticklebacks, and minnows. The Water-frog Bit, *Hydrocharis morsus ranæ*, has heart-shaped leaves, of a bronze colour, beautifully veined, and white flowers, reminding us of the arrow-head in form. The Water-soldier,

GREAT HAIRY WILLOW-HERB.

or Aloe, *Stratiotes aloides*, another very beautiful plant, we likewise found. It is distinguished by its serrated leaves. One of the rarest prizes I ever obtained amongst water plants was the Bladder Wort, *Utricularia vulgaris.* This is a floating plant, which derives its buoyancy from numerous bladders of air attached to the leaves, which only suspend it for the purpose of flowering. Ere it ripens seed the bladders fill with water, and the whole plant sinks to the bottom. Among the interesting plants was the Water

WATER STAR-WORT (*Callitriche verna*).

Star-wort, *Callitriche verna.* The month of June affords the largest number of flowering plants, as well as the largest number of insects, in my experience. I cannot speak so positively about the microscopic world, for then my attention was so occupied with a variety of large forms, that I did not seem to have time for that close attention which I had given in the winter to the minute forms. Hovering over the Lewes brooks, and lighting on the plants to devour the gnats they had captured on the wing, were multitudes of dragon-flies. Of these I identified in one year fourteen species, but some of these were autumnal.

YELLOW RATTLE (*Rhinanthus*).

In the early spring we captured them in the water-net in the larva state, and I once succeeded in rearing in a small aquarium a species *Libellula quadrimaculatum* through its perfect metamorphosis. This species

BUR-REED (*Sparganium Ramosum*).

is distinguished for the beautiful spots on its wings. In the Lewes ponds I obtained several species of fresh-water sponges, and the White Water-lily, *Nymphæa alba*, and that most exquisite of English water-plants, the Yellow Water-lily, *Nuphar lutea*, known from its

GIPSY-WORT (*Lycopus*).

peculiar smell as the brandy bottle. The leaves of these lilies were covered with the fry of *Limnea planorbis*, and the little *Neritina fluviatilis*, and little fresh-water limpets, *Ancylus fluviatilis*, adhered to them also. The *Vitrina pellucida* I once found on the Yellow Iris. The Water Persicaria, *Polygonum amphibium*, noticeable from its bright pink flowers, was

ŒNANTHE CROCATA.

likewise common in a ditch adjoining the Ouse. The Hairy Willow-herb, *Epilobium hirsutum*, was wildly luxuriant, and at the end of the month the Rattle and the Bur-reed, *Sparganium ramosum*, were very abundant, as also the Gipsy Wort, *Lycopus europæus*. That poisonous plant, the Water Drop-wort, *Œnanthe crocata*, was very common on Ouse banks.

MOONWORT (*Botrychium lunaria*).

Few pursuits are more pleasing than to sit upon a river's bank and watch the varied animal and plant-life under the shade of some large tree. If a rod is at hand, and fishing can be combined

ROUND-LEAVED SUNDEW (*Drosera rotundifolia*).

with contemplation, the pleasure is enhanced to most of us.

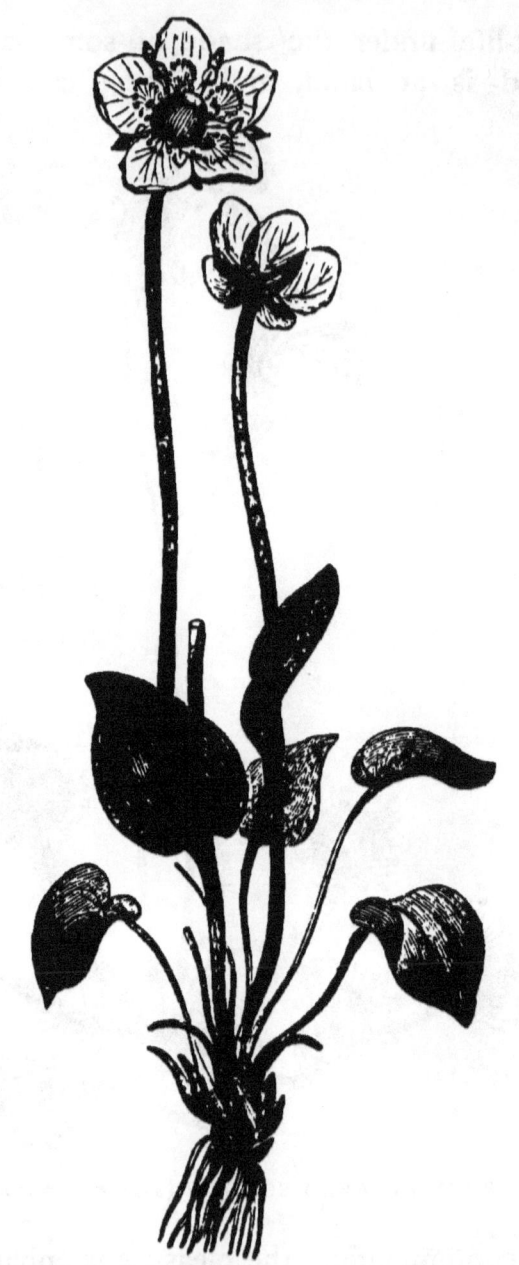

GRASS OF PARNASSUS (*Parnassus palustris*).

Birds'-nesting is one of the most exciting pursuits, particularly in marshes, where the most unexpected surprises occur to those who have the power to keep still and watch. Among the birds which we

MEADOW CRANE'S-BILL (*Geranium pratense*).

frequently noticed in Sussex were the Wagtails. The nests of these species are rather difficult to discover; we found those of two, the Pied and Ray's, along the banks of the Sussex Ouse. The Wood Betony (*Betonica officinalis*) was extremely luxuriant along the Ouse banks. It is a plant which can accommodate itself to much drier situations. On the elevated downs of Sussex occurs the Moonwort. I once found it a few miles from Lewes in a hilly bog, in company with the round-leaved Sundew (*Drosera rotundifolia*), that interesting plant which yields the homeopathist a useful remedy for throat diseases, and by its peculiar habit of catching and digesting flies supplies the evolutionist with material for speculation. The Grass of Parnassus (*Parnassus palustris*) is likewise a Sussex marsh plant, although I never found it at Lewes.

The Bulrush (*Typha latifolia*) I found plentifully in Sussex. Along the banks of most rivers in the south of England is to be found the Meadow Crane's-Bill (*Geranium pratense*), a plant of exquisite purple flowers, surpassing in beauty those of any British species.

In the month of July the growth of plants is so luxuriant as greatly to obscure animal life; hence in marshy districts, in my experience, it is not easy to find a variety of species, as in June; still many species are only to be found in this month. The moths affecting water-plants, notably the genus *Hydrocampa*, a family whose caterpillars live in the water, are beautiful, silvery creatures with porcelain-like marks of brown on them. Then there are the numerous reed moths, and the large Copper (*Chrysophanus dispar*), now extinct in the fenny districts of Cambridgeshire, where it was once common. The numerous neuropterous insects, and the still more numerous Diptera and Coleoptera love the water; Hymenoptera as an order avoid it, having nothing to do with it in their larval condition.

The marsh plants of the rest of the year I have treated in other chapters, so will close my remarks by suggesting that no better place for spring observation and collecting can be found than the neighbourhood of Lewes.

THE END

WYMAN AND SONS, PRINTERS,
GREAT QUEEN STREET, LINCOLN'S INN FIELDS
LONDON, W.C.

Society for Promoting Christian Knowledge.

NATURAL HISTORY RAMBLES.

Fcap. 8vo., Cloth boards, 2s. 6d. each.

LANE AND FIELD.

By the Rev. J. G. Wood, M.A., Author of "Homes without Hands," &c., &c.

LAKES AND RIVERS.

By C. O. Groom Napier, F.G.S., Author of "The Food, Use, and Beauty of British Birds," &c., &c.

MOUNTAIN AND MOOR.

By J. E. Taylor, F.L.S., F.G.S., Editor of "Science-Gossip."

THE SEA-SHORE.

By Professor P. Martin Duncan, M.B. (London), F.R.S., Honorary Fellow of King's College, London.

THE WOODLANDS.

By M. C. Cooke, M.A., LL.D.

UNDERGROUND.

By J. E. Taylor, F.L.S., F.G.S., Editor of "Science-Gossip."

	s.	d.

Beauty in Common Things. Illustrated by 12 Drawings from Nature, by Mrs. J. W. Whymper, and printed in Colours, with descriptions by the Author of "Life Underground," &c. 4to. ....*Cloth boards* 10 6

Botanical Rambles. By the late Rev. C. A. JOHNS, B.A., F.L.S. With illustrations and woodcuts. Royal 16mo..*Cloth boards* 2 0

Flowers of the Field. By the late Rev. C. A. JOHNS, B.A., F.L.S. With numerous woodcuts. Fcap. 8vo... *Cloth boards* 5 0

Wild Flowers. By ANNE PRATT, Author of "Our Native Songsters," &c. With 192 coloured plates, in two volumes. 16mo. ....*Cloth boards* 16 0

Forest Trees (The) of Great Britain. By the Rev. C. A. JOHNS, B.A., F.L.S. New Edition. With 150 woodcuts. Post 8vo. .... *Cloth boards* 5 0

Natural History of the Bible (The). By the Rev. CANON TRISTRAM, Author of "The Land of Israel," &c. With numerous illustrations. Crown 8vo. *Cloth boards* 7 6

Animal Creation (The). A popular Introduction to Zoology. By THOMAS RYMER JONES, F.R.S. With 488 woodcuts. Post 8vo. .... *Cloth boards* 7 6

Lessons from the Animal World. By CHARLES and SARAH TOMLINSON. With 162 woodcuts, in two volumes. Fcap. 8vo. .... *Cloth boards* 4 0

Birds' Nests and Eggs. With 22 coloured plates of Eggs. Square 16mo. .... *Cloth boards* 3 0

British Birds in their Haunts. By the late Rev. C. A. JOHNS, B.A., F.L.S. With 190 engravings by Wolf and Whymper. Post 8vo. .... *Cloth boards* 10 0

	s.	d.

British Animals. With 12 coloured plates. 16mo. *Ornamental covers* — 1 6

Birds of the Sea-shore. With 12 coloured plates. 16mo. .... *Cloth boards* 1 8

Evenings at the Microscope; or, Researches among the Minuter Organs and Forms of Animal Life. By PHILIP HENRY GOSSE, F.R.S. A new Edition revised and annotated. With 112 woodcuts. Post 8vo. *Cloth boards* 4 0

Familiar History of British Fishes. By FRANK BUCKLAND, Inspector of Salmon Fisheries for England and Wales. With a Frontispiece and 134 woodcuts. Crown 8vo. .... *Cloth boards* 5 0

General Knowledge (First Steps in). By Mrs. C. TOMLINSON.

   Part I. THE STARRY HEAVENS, demy 16mo, with several diagrams, &c. .... *cloth boards* 1 0

       II. THE SURFACE OF THE EARTH, demy 16mo, with several wood-cuts ...... *cloth boards* 1 0

       III. THE ANIMAL KINGDOM, demy 16mo, with several wood-cuts ....*cloth boards* 1 0

       IV. THE VEGETABLE KINGDOM, demy 16mo, with several wood-cuts ...... *cloth boards* 1 0

       V. THE MINERAL KINGDOM, demy 16mo, with several wood-cuts .... *cloth boards* 1 0

Natural History (Illustrated Sketches of): consisting of Descriptions and Engravings of Animals. With numerous woodcuts, in 2 vols. Fcap. 8vo. Series I. and II. .... *Cloth boards, each Vol.* 2 6

Our Native Songsters. By ANNE PRATT, Author of "Wild Flowers." With 72 coloured plates. 16mo. *Cloth boards* 8 0

	s.	d.
Selborne (The Natural History of). By the Rev. GILBERT WHITE. With Frontispiece, Map, and 50 woodcuts. Post 8vo. ....*Cloth boards*	2	6
Ocean (The). By PHILIP HENRY GOSSE F.R.S., Author of "Evenings at the Microscope." With 51 illustrations and woodcuts. Post 8vo. *Cloth boards*	4	6
Dew-drop and the Mist (The): an Account of the Phenomena and Properties of Atmospheric Vapour in various Parts of the World. By CHARLES TOMLINSON, F.C.S. With woodcuts and diagrams. Fcap. 8vo. *Cloth boards*	2	6
Frozen Stream (The): an Account of the Formation and Properties of Ice in various Parts of the World. By CHARLES TOMLINSON. With woodcuts and diagrams. Fcap. 8vo. ....*Cloth boards*	1	6
Rain-Cloud and Snow-Storm: an Account of the Nature, Formation, Properties, Dangers, and Uses of Rain and Snow. By C. TOMLINSON. With numerous woodcuts and diagrams. Fcap. 8vo. *Cloth boards*	2	6
Tempest (The): an Account of the Origin and Phenomena of Wind in various Parts of the World. By CHARLES TOMLINSON. With numerous woodcuts and diagrams. Fcap. 8vo..............................*Cloth boards*	2	6
Thunder-Storm (The): an Account of the Properties of Lightning and of Atmospheric Electricity in various Parts of the World. By CHARLES TOMLINSON. With numerous woodcuts and diagrams. Fcap. 8vo. *Cloth boards*	2	6
Winter in the Arctic Regions and Summer in the Antarctic Regions. By CHARLES TOMLINSON. With two maps, and several illustrations and woodcuts. Crown 8vo. ....*Cloth boards*	4	0

THE HOME LIBRARY.

Crown 8vo. Cloth Boards, 3s. 6d. each.

The House of God, the Home of Man. By the Rev. G. E. JELF, M.A., Vicar of Saffron Walden.

The Inner Life, as Revealed in the Correspond-ence of Celebrated Christians. Edited by the late Rev. T. ERSKINE.

Savonarola: his Life and Times. By the Rev. WILLIAM R. CLARK, M.A., Author of "The Comforter," &c.

NON-CHRISTIAN RELIGIOUS SYSTEMS.

Fcap. 8vo., with Map. Cloth Boards, 2s. 6d. each.

Buddhism. Being a Sketch of the Life and Teachings of Gautama, the Buddha. By T. W. RHYS DAVIDS, of the Middle Temple.

The Corân: Its Composition and Teaching, and the Testimony it bears to Holy Scriptures. By Sir WILLIAM MUIR, K.C.S.I., LL.D.

Hinduism. By MONIER WILLIAMS, M.A., D.C.L., &c.

Islam and its Founder. By J. W. H. STOBART, B.A., Principal, La Martinière College, Lucknow.

SPECIFIC SUBJECTS.

NEW EDUCATIONAL CODE.

Fcap. 8vo., with Diagrams. Limp Cloth, 4d. each.

Algebra. By W. H. H. HUDSON, M.A., Mathematical Lecturer, and late Fellow of St. John's College, Cambridge.

(Answers to the Examples in the above, limp cloth, **8d.**)

Euclid. Books 1 and 2. Edited by W. H. H. HUDSON, M.A., late Fellow of St. John's College, Cambridge.

Elementary Mechanics. By W. GARNETT, M.A., Fellow of St. John's College, Cambridge.

Physical Geography. By the Rev. T. G. BONNEY, M.A., F.G.S., &c., late Tutor of St. John's College, Cambridge.

CONVERSION OF THE WEST.

Fcap. 8vo. Cloth Boards, 2s. each.

The Continental Teutons. By the Very Rev. CHARLES MERIVALE, D.D., D.C.L., Dean of Ely. With Map.

The Celts. By the Rev. G. F. MACLEAR, D.D., Headmaster of King's College Schools. With Two Maps.

The English. By the same Author. With Two Maps.

The Northmen. By the same Author. With Map.

The Slavs. By the same Author. With Map.

THE FATHERS FOR ENGLISH READERS.

Fcap. 8vo. Cloth Boards, 2s. each.

Apostolic Fathers (The). By the Rev. H. S. HOLLAND, Student of Christ Church, Oxford.

Defenders of the Faith (The); or, the Christian Apologists of the Second and Third Centuries. By the Rev. F. WATSON, M.A., Rector of Starston, Norfolk.

Saint Augustine. By the Rev. WILLIAM R. CLARK, M.A., Vicar of Taunton, Author of "The Comforter."

Saint Jerome. By the Rev. EDWARD L. CUTTS, D.D., Author of "Turning Points of Church History," &c.

ANCIENT HISTORY FROM THE MONUMENTS.

Fcap. 8vo., with Illustrations. Cloth Boards, 2s. each.

- **Assyria, from the Earliest Times to the Fall of** Nineveh. By the late GEORGE SMITH, Esq., of the Department of Oriental Antiquities, British Museum.
- **Babylonia, The History of.** By the late GEORGE SMITH, Esq. Edited by the Rev. A. H. SAYCE, Assistant Professor of Comparative Philology, Oxford.
- **Egypt, from the Earliest Times to B.C. 300.** By S. BIRCH, LL.D., &c.
- **Persia, from the Earliest Period to the Arab** Conquest. By W. S. W. VAUX, M.A., F.R.S.
- **Greek Cities and Islands of Asia Minor.** By W. S. W. VAUX, M.A., F.R.S.

THE HEATHEN WORLD AND ST. PAUL.

Fcap. 8vo., with Map. Cloth Boards, 2s. each.

- **St. Paul in Damascus and Arabia.** By the Rev. GEORGE RAWLINSON, M.A., Canon of Canterbury, Camden Professor of Ancient History, Oxford.
- **St. Paul in Greece.** By the Rev. G. S. Davies, M.A., Charterhouse, Godalming.
- **St. Paul at Rome.** By the Very Rev. CHARLES MERIVALE, D.D., D.C.L., Dean of Ely.
- **St. Paul in Asia Minor, and at the Syrian Antioch.** By the Rev. E. H. PLUMPTRE, D.D., Prebendary of St. Paul's, Vicar of Bickley, Kent.

MANUALS OF ELEMENTARY SCIENCE.

Fcap. 8vo., with Illustrations. Limp Cloth, 1s. each.

Physiology. By F. LE GROS CLARKE, F.R.S., St. Thomas's Hospital.

Geology. By the Rev. T. G. BONNEY, M.A., F.G.S., Fellow and Tutor of St. John's College, Cambridge.

Chemistry. By ALBERT J. BERNAYS, Professor of Chemistry, St. Thomas's Hospital.

Astronomy. By W. H. CHRISTIE, M.A., Trinity College, Cambridge; the Royal Observatory, Greenwich.

Botany. By ROBERT BENTLEY, Professor of Botany in King's College, London.

Zoology. By A. NEWTON, M.A., F.R.S., Professor of Zoology and Comparative Anatomy in the Univ. of Cambridge.

Matter and Motion. By J. CLERK MAXWELL, M.A., Trin. Coll., Camb., Univ. Professor of Experimental Physics.

Spectroscope, The Work of the. By RICHARD A. PROCTOR, Esq.

MANUALS OF HEALTH.

Fcap. 8vo. Limp Cloth, 1s. each.

Health and Occupation. By BENJAMIN WARD RICHARDSON, Esq., F.R.S., M.D., &c.

On Personal Care of Health. By the late E. A. PARKES, M.D., F.R.S., Professor of Military Hygiène in the Army Medical Schools, Netley.

Food. By ALBERT J. BERNAYS, Professor of Chemistry at St. Thomas's Hospital.

Water, Air, and Disinfectants. By W. NOEL HARTLEY, Esq., King's College.

Depositories:
77, GREAT QUEEN STREET, LINCOLN'S-INN FIELDS;
4, ROYAL EXCHANGE; AND 48, PICCADILLY.

www.ingramcontent.com/pod-product-compliance
Lightning Source LLC
Chambersburg PA
CBHW021408230426
43666CB00006B/671